The Eye in Chromosome Duplications and Deficiencies

OPHTHALMOLOGY SERIES

Editor: **Paul Henkind, M.D., Ph.D.**
Professor and Chairman
Department of Ophthalmology
Albert Einstein College of Medicine
Montefiore Hospital
New York, New York

Other volumes in preparation

The Eye in Chromosome Duplications and Deficiencies

MARCELLE JAY

Department of Experimental Ophthalmology
Institute of Ophthalmology
University of London
London, England

MARCEL DEKKER, INC. New York and Basel

Library of Congress Cataloging in Publication Data

Jay, Marcelle.

The eye in chromosome duplications and deficiencies

(Ophthalmology series ; v. 2)
Includes bibliographical references and indexes.
1. Eye--Abnormalities--Genetic aspects. 2. Human chromosome abnormalities. 3. Ocular manifestations of general diseases. I. Title. II. Series. [DNLM: 1. Eye--Abnormalities. 2. Chromosome abnormalities. 3. Chromosome aberrations. W1 OP373 v. 2 / WW101 J42e]
RE906.J38 617.7'042 76-1796
ISBN 0-8247-6455-2

MARCEL DEKKER, INC.

270 Madison Avenue, New York, New York 10016

Current printing (last digit):
10 9 8 7 6 5 4 3 2 1

PRINTED IN THE UNITED STATES OF AMERICA

FOREWORD

There has been, in recent years, an explosion of information dealing with genetics and ophthalmology. Much of the information does not appear in the ophthalmologic literature but rather has been concentrated in journals dealing with heredity and genetics. Marcelle Jay has had a long interest in genetics and, because of her particular association with ophthalmology, has coupled her interest so that she has developed an outstanding knowledge of genetic diseases as they affect the eye. In this volume, which is an outgrowth of her thesis, she has compiled all of the extant information on the relationship of chromosomal duplications and deficiencies upon the human eye. She provides outstanding source material for anyone who would desire to pursue the subject in depth. Clinicians in the area of ophthalmology, pediatrics, and otolaryngology will find this to be a handy reference volume. The enormous task in compiling all the material is one that would have caused most researchers to shy away, and we can only admire such an effort, which makes our work easier.

Paul Henkind, M.D., Ph.D.

PREFACE

The study of human cytogenetics has made very rapid advances in the past fifteen years due to improved techniques, and new syndromes of chromosomal abnormalities are still being discovered.

The purpose of this book is to review the literature on ocular abnormalities associated with structurally abnormal chromosomes, deletions, and duplications. These so-called "new chromosomal syndromes" have not been previously discussed in detail, no doubt because it is only recently that technical advances have made their detection possible.

The literature on the Turner syndrome has been included, because although the ocular abnormalities in this syndrome are well known, and it is not a deletion syndrome, it has a relatively high incidence and it would be useful for ophthalmologists to have a recent review for reference.

The greater part of the review of the literature is devoted to deletions rather than to duplications. There are earlier references in the literature to abnormal chromosomes, but these have been omitted wherever the origin of the chromosomal segment in excess was unknown. Most of the references given in the chapters on duplications are relatively recent; this is probably because the newer banding techniques have made possible the accurate identification of a given chromosomal segment. It is inevitable that such a review cannot be exhaustive, but I hope that all the more important publications have been included, and the references are complete (except for human error) to the end of 1974.

It is obviously invidious to isolate the ocular from other congenital abnormalities which occur in deletion and duplication syndromes, and also to isolate any one sign. For this reason, the whole phenotype has been given wherever possible, with particular emphasis on facial abnormalities. The abnormality of the karyotype and case numbers are given as in the original papers.

It is interesting to note that the majority of cases have been reported by physicians and geneticists, and only a few by ophthalmologists, and yet some ocular abnormality has been noted in almost every case. It would seem that a closer study of these syndromes by ophthalmologists might prove fruitful, and stricter criteria might be applied for the evaluation of certain signs. For example, hypertelorism is reported in many cases, whereas the more correct term "telecanthus" is very rarely used and then nearly always when an ophthalmologist has examined the case.

A brief introduction to cytogenetics has been given in the hope that this review will thereby be more meaningful when read by those not readily acquainted with the terms and techniques which are now in use, and will enable them to keep up with recent advances.

Anyone writing about ophthalmology and cytogenetics must acknowledge the inspiration given by the work of François, Berger, and Saraux on this topic; their book, "Les Aberrations Chromosomiques en Ophtalmologie," Paris, Masson & Cie, 1972, is widely quoted throughout this review.

I would like to thank my teachers in genetics and in ophthalmology, Professor Cedric Carter, Dr. Rufus Howard, and Mr. Barrie Jay for their patience and their valuable comments. My thanks are also due to the Department of Audio-Visual Communications at the Institute of Ophthalmology, London; in particular to Mr. Terry Tarrant for his line drawings, and to Mr. Roger Fletcher for preparing photographs. Dr. Michael Daker prepared the karyotypes and I am grateful to Professor Paul Polani of the Paediatric Research Unit at Guy's Hospital Medical School, London, for permission to publish Figures 1-7, 1-9, 1-10, 1-11, 1-12, and 1-16 which have appeared in "Handbuch der Urologie," Vol. 15: Urology in Childhood, by Innes Williams, Heidelberg, Springer Verlag, 1974, and also to Dr. Jack Singer of the Paediatric Research Unit for his permission to use Figure 1-8 which is reproduced from "The Fetus: Physiology and Medicine," edited by R. W. Beard and P. W. Nathanielsz, London, W. B. Saunders, Ltd., 1976. I would also like to thank Dr. M.-O. Réthoré of the Hôpital des Enfants Malades, Paris, for permission to use Figure 10-1 which appeared as Figure 2 in the article "Trisomie 9p par t(4;9)(q34;q21)mat" by M.-O. Réthoré, J. Ferrand, B. Dutrillaux, and J. Lejeune, published in Annales de Génétique, 17: 157-161 (1974).

Several authors have been kind enough to let me use previously unpublished photographs for which they own the copyright and I am most grateful to Dr. Renata Lax and Dr. Michael Ridler of the Kennedy Galton Institute, Harperbury Hospital, Radlett, Herts., to Mr. L. J. Butler of Queen Elizabeth Hospital for Children, Hackney Road, London E2 8PS, to Mr. A. J. Bron of the Nuffield Laboratory of Ophthalmology, Walton

Street, Oxford, and to Professor Norman Ashton of the Institute of Ophthalmology, Judd Street, London, WC1H 9QS, for their help and beautiful photographs.

Finally I am most grateful to Miss Audrey Jones who typed the manuscript in its present form.

Marcelle Jay
London, England

CONTENTS

FOREWORD — Paul Henkind, M. D., Ph. D. iii
PREFACE v
KEY TO TABLES xi

1. INTRODUCTION TO CYTOGENETICS 1

Cell Division 1
Examination of Chromosomes 5
Classification and Nomenclature 6
Identification of Chromosomes 10
Numerical Changes in the Karotype 17
Structural Changes in the Karotype 17
The Sex Chromosomes 24
Suggested Readings 26

2. INCIDENCE OF CHROMOSOME ABNORMALITIES 29

Spontaneous Abortions 29
Perinatal Deaths 29
Newborn Infants 30
References 32

3. DEFICIENCIES OF GROUP B CHROMOSOMES 33

Introduction 33
4p- or Wolf-Hirschhorn Syndrome 33
5p- or Cri-du-Chat (Cat Cry) Syndrome 44
References 63

4. DEFICIENCIES OF GROUP C CHROMOSOMES 73

Ocular Abnormalities 73
References 75

5. DEFICIENCIES OF GROUP D CHROMOSOMES 77

Group D Rings 77
Group D Deletions 85
References 95

6. DEFICIENCIES OF GROUP E CHROMOSOMES 101

Deletion of the Short Arm of Chromosome 18 (18p-) 101
Deletion of the Long Arm of Chromosome 18 (18q-) 110
Ring Chromosome 18 (18r) 119
References 127

7. DEFICIENCIES OF GROUP F CHROMOSOMES 137

Ocular Abnormalities 137
References 137

8. DEFICIENCIES OF GROUP G CHROMOSOMES 139

Ocular Abnormalities 139
References 145

9. MONOSOMY X 149

Turner Syndrome and Turner Mosaics 149
References 163

10. DUPLICATIONS 169

Partial Trisomies 169
Summary 176
Pericentric Inversions 189
References 192

11. BALANCED TRANSLOCATIONS 201

Ocular Abnormalities 201
References 207

12. ANEUPLOID TRANSLOCATIONS 211

Ocular Abnormalities 211
References 212

EPILOGUE 213
GLOSSARY 215
AUTHOR INDEX 221
SUBJECT INDEX 243

KEY TO TABLES

+	Sign present bilaterally
(+)	Sign present unilaterally
0	Sign absent
?	Sign doubtful
L	Left
R	Right
Con	Convergent
Div	Divergent
Alt	Alternating
M	Myopia
A	Astigmatism
D	Deuteranopia
P	Protanopia

Wherever possible, the case numbers used and the chromosomal abnormalities are those given in the original papers.

CHAPTER 1

INTRODUCTION TO CYTOGENETICS

I. CELL DIVISION

The human cell passes through several stages during its life-span. The first stage is interphase, during which the cell is at rest and the chromosomes appear as single threads. These chromosomes are deeply-staining structures in the nuclei of cells and are composed of deoxyribonucleic acid (DNA) combined with a protein. Towards the end of interphase there is the S period, during which DNA is synthesized and replication occurs. The chromosomes then double and appear as two identical threads or chromatids which are joined by a main constriction, the centromere (Fig. 1-1). The nuclei of all human somatic cells contain 46 chromosomes, consisting of 22 identical pairs of chromosomes or homologues and two sex chromosomes: a pair of X chromosomes in the female and an X and Y chromosome in the male. Somatic cells divide by the process of mitosis, whereby each daughter cell retains the same number of chromosomes as the parent cell.

FIG. 1-1. Diagrammatic representation of a metaphase chromosome.

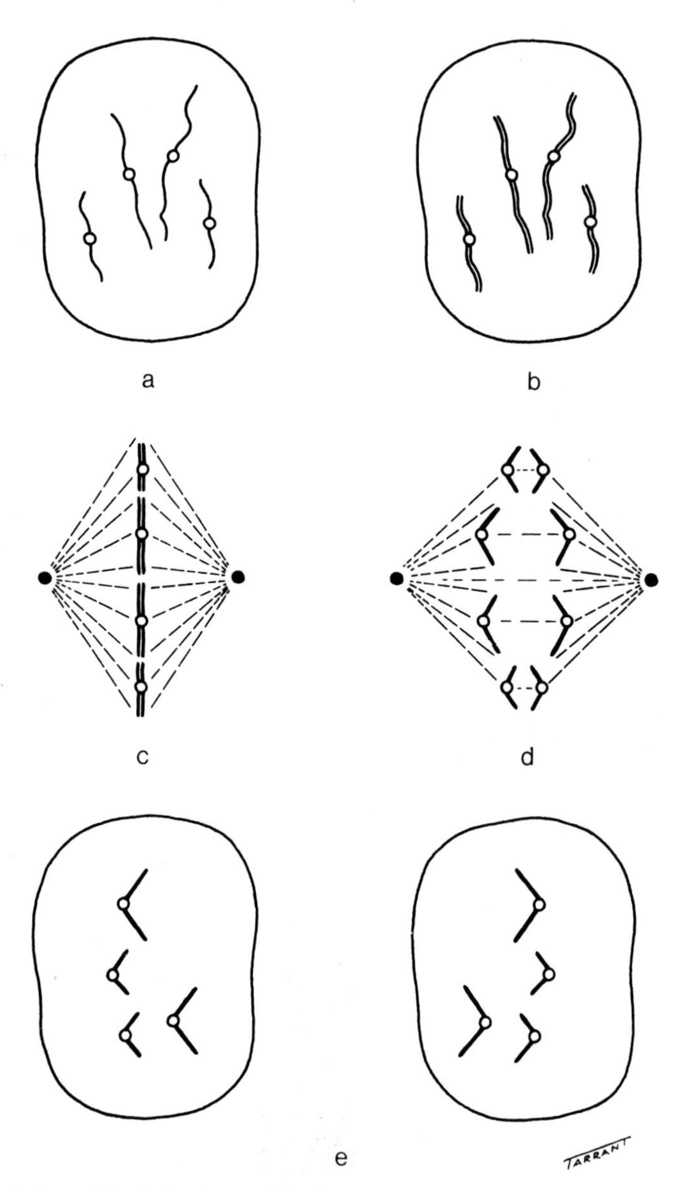

FIG. 1-2. The behaviour of 4 chromosomes at mitosis. (a) Interphase with 4 chromosomes in the parent cell. (b) Prophase during which each chromosome is doubled into identical chromatids. (c) Formation of the spindle at metaphase. (d) Each chromosome splits during anaphase. (e) Telophase and the formation of two daughter cells, each with 4 chromosomes.

A. Mitosis

Mitosis follows interphase and is divided into four stages: prophase, metaphase, anaphase, and telophase (Fig. 1-2). Each stage is defined as follows:

1. Prophase. During this stage the chromosomes contract, each chromatid becomes tightly coiled, and the nuclear membrane disappears.
2. Metaphase. It is at this stage that chromosomes can be observed readily by using various staining techniques, which are described later in this chapter. A spindle is formed by long strands of fibres which pass between the two poles of the cell. The chromosomes then move so as to lie in an equatorial plane equidistant from the two poles of the spindle, where they are distributed evenly.
3. Anaphase. The centromeres split longitudinally and the single chromatids then move to the opposite poles of the spindle.
4. Telophase. In this final stage, the spindle disappears and a new cell membrane is formed. The two daughter cells separate and the chromosomes again become thin and threadlike. The cell now resumes its resting interphase stage.

B. Meiosis

The process whereby gametes (sperm and ova) are formed is called meiosis. This is a form of cell division resulting in the number of chromosomes being halved. Each human gamete contains 23 chromosomes, the haploid (single) number, while each human somatic cell contains 46 chromosomes, the diploid (double) number. When two haploid gametes fuse at fertilization, the resultant cell is the zygote which has a diploid number of chromosomes.

Meiosis consists of two divisions, which are each divided into prophase, metaphase, anaphase, and telophase (Fig. 1-3).

1. Prophase I. During this long stage, the chromosomes, consisting of two chromatids, form pairs which are called bivalents. Each bivalent consists of two homologous chromosomes, one maternal and the other paternal in origin. The exchange of genetic material that may take place at this time (Fig. 1-4) is called crossing-over and the points at which it

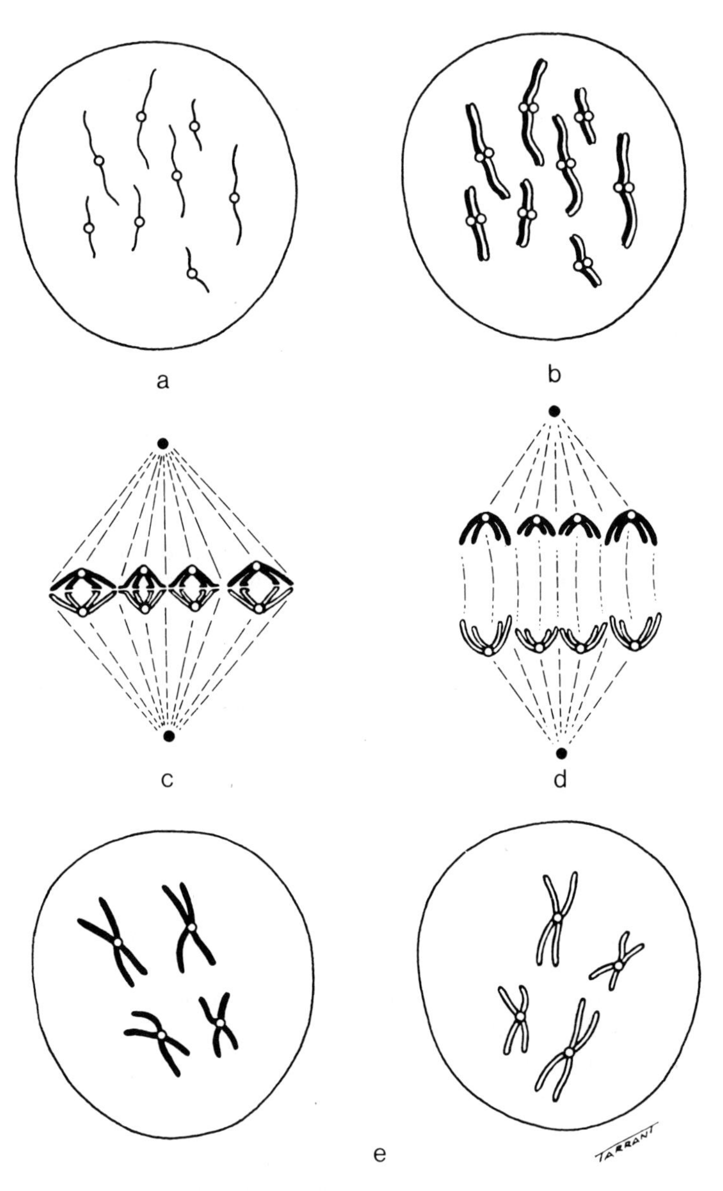

FIG. 1-3. First division of meiosis showing the behaviour of 8 chromosomes. (a) Interphase with 8 chromosomes in the parent cell. (b) Prophase 1 and the formation of bivalents. (c) Metaphase 1 during which the bivalents lie on the equatorial plane. (d) Anaphase 1, the bivalents divide. (e) Telophase 1, two daughter cells are formed, each with 4 chromosomes. These cells then proceed to the second division of meiosis which occurs in the same way as mitosis.

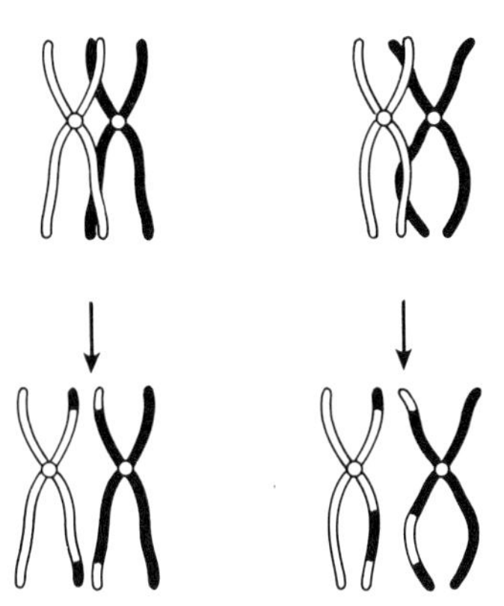

FIG. 1-4. Crossing-over. The exchange of genetic material between chromosomes of maternal and paternal origin.

occurs are called the chiasmata. During prophase I the bivalents contract, and at the end of this stage they reach maximal condensation. The bivalents that are joined only at the chiasmata now tend to fall apart.

2. Metaphase I. The nuclear membrane disappears, the spindle forms, and the bivalents move to the equator of the spindle, with the centromeres of each bivalent lying on either side of the equator.

3. Anaphase I. The bivalents split and then the chromosomes move to opposite poles of the cell so that half their number is at one pole and half at the opposite pole.

4. Telophase I. The haploid groups of chromosomes may uncoil and a new membrane may form, or they may proceed to the second division in meiosis. This consists of prophase II, metaphase II, anaphase II, and telophase II, and occurs exactly as in mitosis, but only the haploid number of chromosomes is involved.

II. EXAMINATION OF CHROMOSOMES

The analysis of chromosomes is made on cells which are dividing, since chromosomes can only be examined in detail during the metaphase stage of mitosis, or during the late prophase and metaphase stages of meiosis. Various tissues are available for culture, but those used are usually the lymphocytes of peripheral blood and the fibroblasts of skin. The culture of lymphocytes is a relatively simple procedure and has the advantage of requiring only two to three days. The culture of fibroblasts is more elaborate and expensive, requiring 14 days, but it has the advantage that the culture can be prolonged and also that the cells can be deep frozen. This method is used for the culture of cells in the amniotic fluid which is obtained by amniocentesis.

Another method of chromosome analysis is the direct study of bone-marrow cells, as used in the chromosome study of patients with leukemia. Since bone-marrow cells are already dividing in vivo, the time required for analysis is much shorter as compared with previous methods. Using a recent modification of this method, a chromosome count may be obtained in two hours which is of great assistance to physicians faced with problems of differential diagnosis and management. In certain cases, testicular biopsy is performed to obtain cells in meiosis; this procedure is sometimes used in cases of infertility. Chromosomes in meiosis have their own characteristic appearance, but the interpretation of the results obtained requires great skill. Other tissues are also used in experimental work, as for example, the cornea in mammals.

III. CLASSIFICATION AND NOMENCLATURE

The appearance of chromosomes at metaphase shows two identical chromatids joined by the centromere (Fig. 1-1). Various classifications can be made according to size and structure, with one classification depending upon the relative position of the centromere (Fig. 1-5). A chromosome is said to be (a) metacentric when the centromere is at its centre; (b) it is submetacentric when the centromere is to one side of the centre; and (c) it is acrocentric when the centromere is towards one end of the chromosome. The centromere divides the chromosome into unequal arms, the short arm being denoted by the letter "p" and the long arm by the letter "q."

A. Chicago Classification

A standard international system of nomenclature was adopted after the Chicago Conference of 1966, whereby chromosomes were divided into groups according to their size and structure. The 22 pairs of autosomes

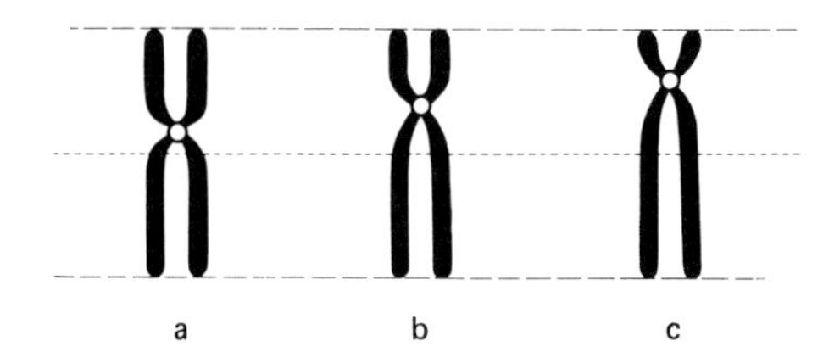

FIG. 1-5. (a) Metacentric chromosome; (b) submetacentric chromosome; and (c) acrocentric chromosome.

were divided in descending order of size into seven groups: A, B, C, D, E, F, and G. The sex chromosomes were X and Y, with the male complement being XY and the female complement XX.

Group A consists of chromosome pairs 1 to 3. These are metacentric and are the largest pairs. Chromosome 3 has arms of equal length.

Group B consists of chromosome pairs 4 and 5 which are submetacentric and long, although shorter than those in group A.

Group C is composed of chromosome pairs 6 to 12 and the X chromosome, and these are all submetacentric.

Group D is formed by the acrocentric chromosome pairs 13 to 15.

Group E consists of chromosome pairs 16 to 18. Of these, pair 16 is metacentric but pairs 17 and 18 are submetacentric.

Group F contains chromosome pairs 19 and 20 which are small and metacentric.

Group G contains chromosome pairs 21 and 22 which are the smallest of the chromosomes and are acrocentric. The Y chromosome also belongs to group G and in most cells it can be distinguished from chromosomes 21 and 22, although its length may vary from one individual to another.

B. Paris Classification

Chromosomes can be further classified by the presence of secondary constrictions and satellites, and by a characteristic linear pattern along the length of the chromosome which can be demonstrated by special techniques. This linear pattern is called banding and the presence of chromosome bands has led to a new system of nomenclature as proposed at the Paris Conference in 1971. A band is a segment of a chromosome which is distinctly lighter or darker than its adjacent segment. The bands are allotted to various regions along the chromosome arms, each region being limited by two adjacent chromosome landmarks such as the centromere or the ends of chromosome arms. Regions and bands are numbered consecutively from the centromere along each chromosome arm. In this way a particular band may be designated by the chromosome number,

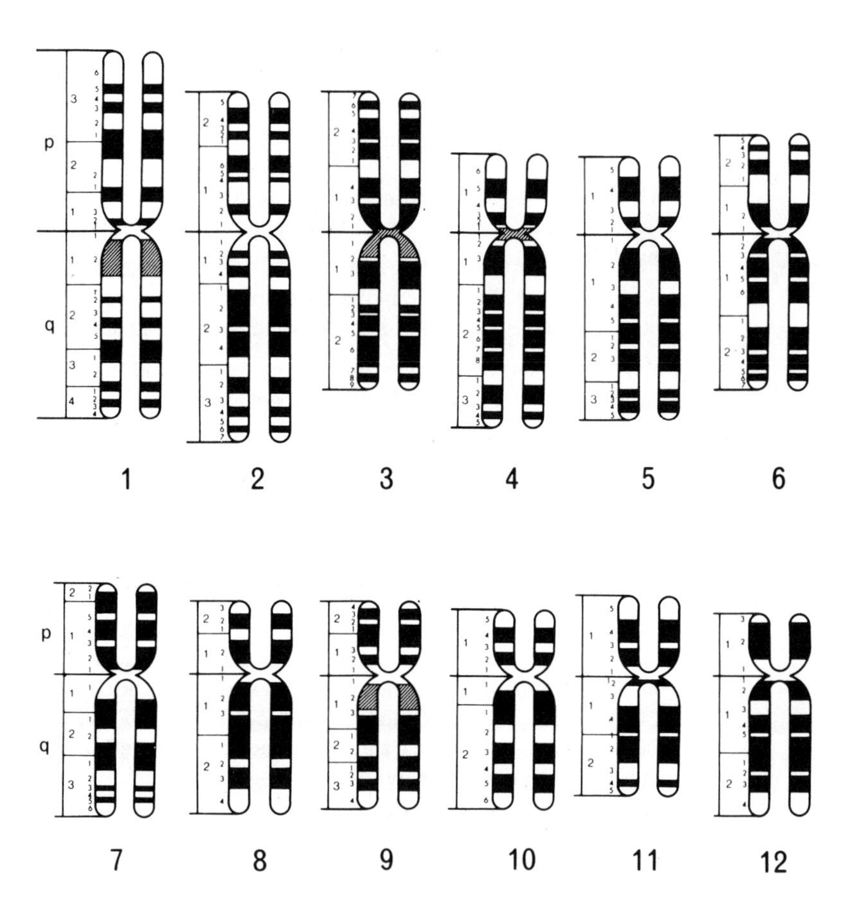

FIG. 1-6. Banding pattern and Paris Classification. (From Paris Conference (1971): Standardization in Human Cytogenetics. In Birth Defects: Orig. Art. Ser. (D. Bergsma, ed.), The National Foundation, March of Dimes, White Plains, N. Y. 8:7(1972).

the arm symbol, the region number, and the band number. For example, 1p33 is chromosome 1, short arm, region 3, band 3. The bands themselves may be further subdivided with the same system of numbering outward from the centromere (Fig. 1-6).

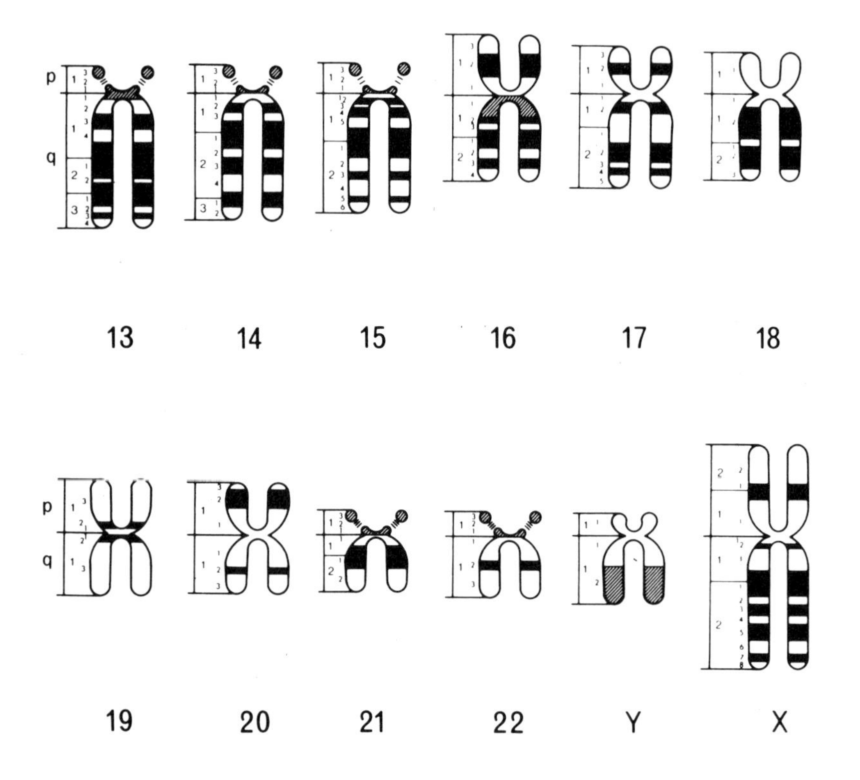

FIG. 1-6 (continued).

IV. IDENTIFICATION OF CHROMOSOMES

There are many different methods of identifying chromosomes, all of which have their own special advantages. The main methods used are autoradiography, fluorescent staining with quinacrine derivatives, and the Giemsa stain and its modifications.

A. Autoradiography

This technique uses "labelling" of chromosomes with a radioactive isotope, tritiated thymidine. Thymidine is a precursor of thymine which is incorporated into the structure of DNA, and the chromosomes take up the labelled thymidine during the S period of DNA synthesis. Tritiated thymidine is added to cells in culture during the S period and labelling is either continuous or pulse. In continuous labelling the cells remain in contact with tritiated thymidine until the preparations are made. In pulse labelling the cells are exposed to tritiated thymidine for a short period, then the excess tritiated thymidine is removed, and the cells continue to grow in culture until the preparations are made. The slides are treated with a special photographic emulsion which records the areas of radioactivity over the chromosomes. These areas appear as black grains and the chromosomes are identified by studying the number and distribution of the grains. The synthesis of DNA is asynchronous in human chromosomes, so that early- or late-replicating chromosomes may be distinguished by the distribution of the grains. Autoradiography is used with success to identify the X chromosome and those of groups B, D, E and G because of their replication patterns, but it is of little use for identifying chromosomes in groups A, C, and F, where different techniques must be used.

B. Fluorescent Studies with Quinacrine Derivatives

Chromosomes produce a specific banding pattern of fluorescence when treated with quinacrine mustard or quinacrine hydrochloride (Fig. 1-7). The resultant intensity of fluorescence can be analysed automatically from photographs and can be plotted on a graph as a function of the length, so that each chromosome has its own specific curve or profile.

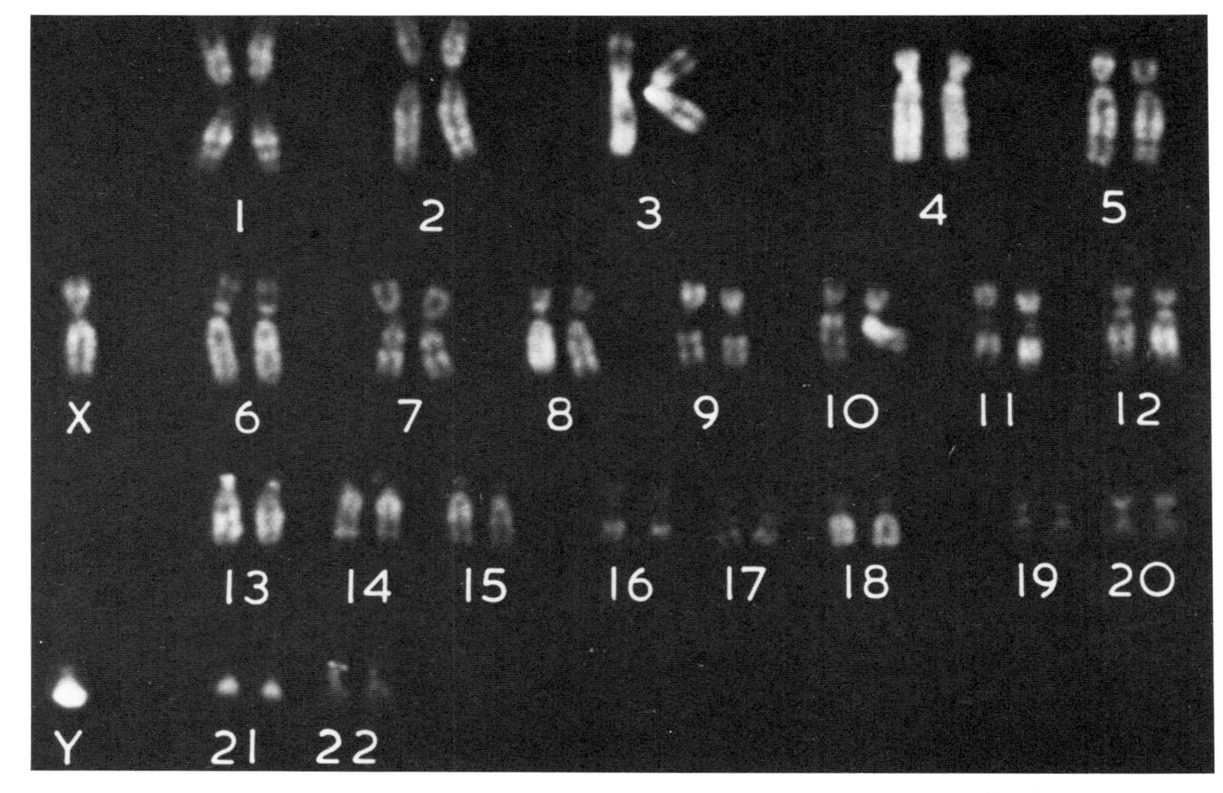

FIG. 1-7. Fluorescent banding pattern, Q bands. (By courtesy of Professor Paul Polani)

This is a refinement of the technique of comparing banding patterns for each chromosome in a qualitative manner.

C. Giemsa Stain

The Giemsa stain will produce a banding pattern which is similar to that obtained by using quinacrine fluorescence (Fig. 1-8). One of the Giemsa stain techniques produces a banding pattern which is the exact reverse of that obtained by other Giemsa staining methods, and is therefore called the reverse Giemsa stain method (Fig. 1-9). The Giemsa stain may be further treated by trypsin, or by controlled heating (denaturation thermique). The resultant bands are called Q-bands, G-bands, or R-bands (Q = quinacrine; G = Giemsa; and R = reverse Giemsa), according to the staining method used. Descriptive diagrams of chromosomes and

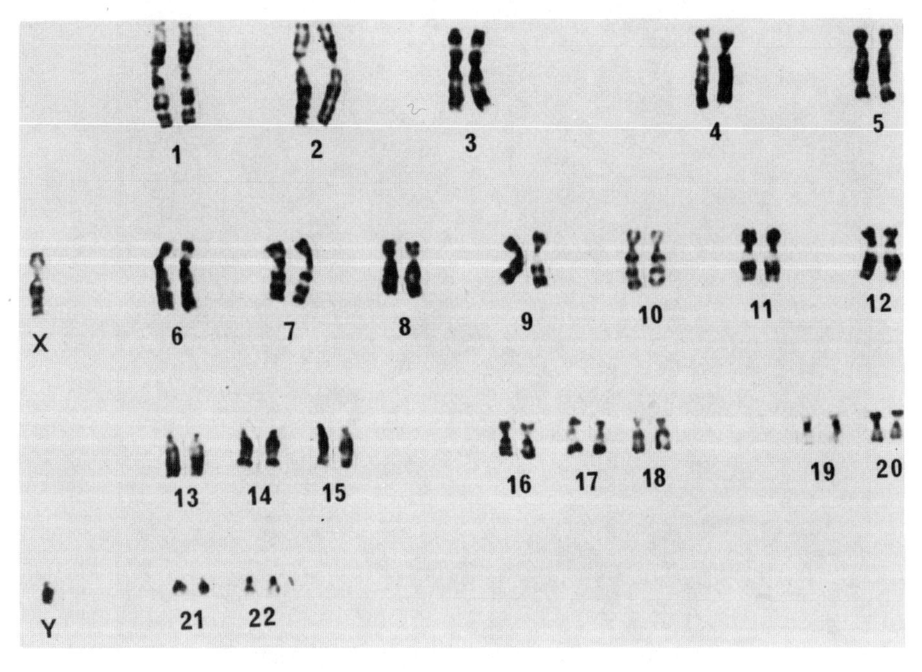

FIG. 1-8. Banding pattern with Giemsa stain, G bands. (By courtesy of Dr. Jack Singer)

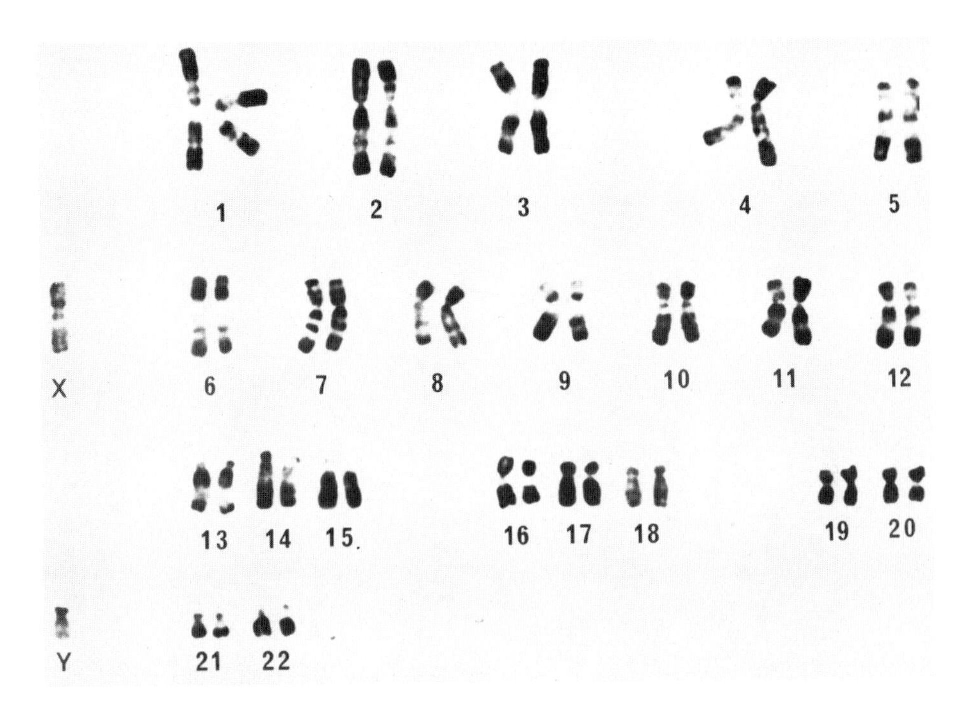

FIG. 1-9. Reverse Giemsa banding pattern, R bands. (By courtesy of Professor Paul Polani)

their banding patterns have recently been published and these include an analysis of R-bands using the Paris nomenclature. Any of these methods or various modifications of them, are used to obtain a karyotype. Figure 1-10 shows a photograph of chromosomes at metaphase in which the chromosomes have been cut out and rearranged according to their group and number. The karyotype is also the formula giving the individual chromosome complement: the normal male karyotype being 46, XY and the normal female karyotype 46, XX (Figs. 1-11 and 1-12).

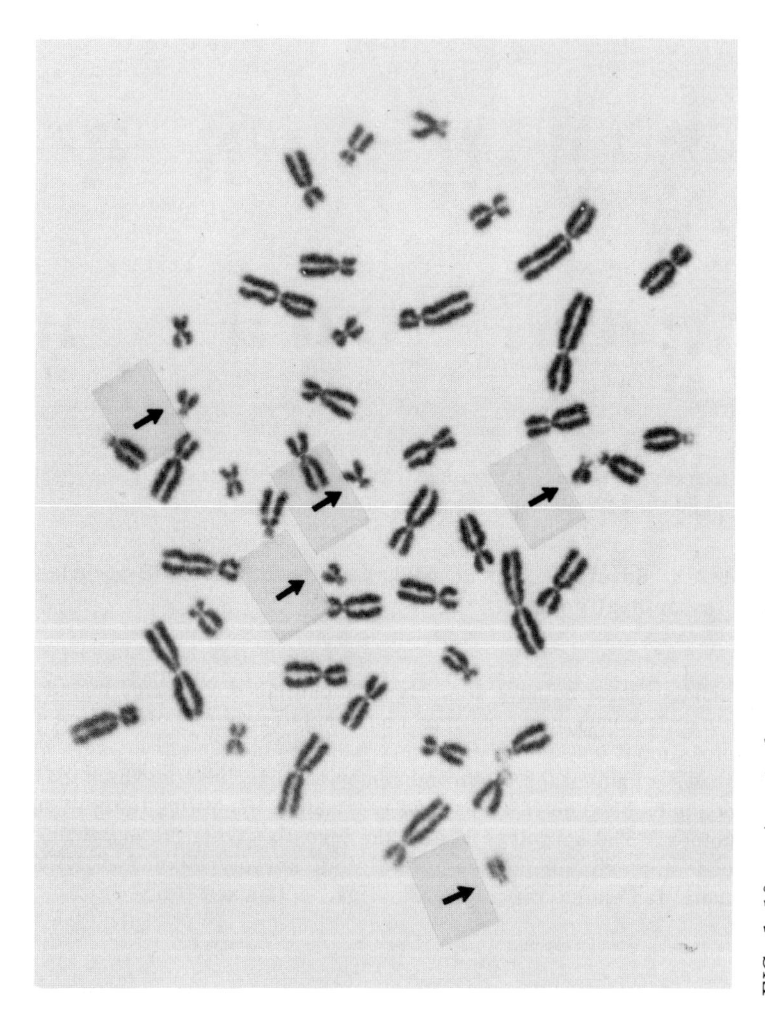

FIG. 1-10. A metaphase spread with 5 G chromosomes. (By courtesy of Professor Paul Polani)

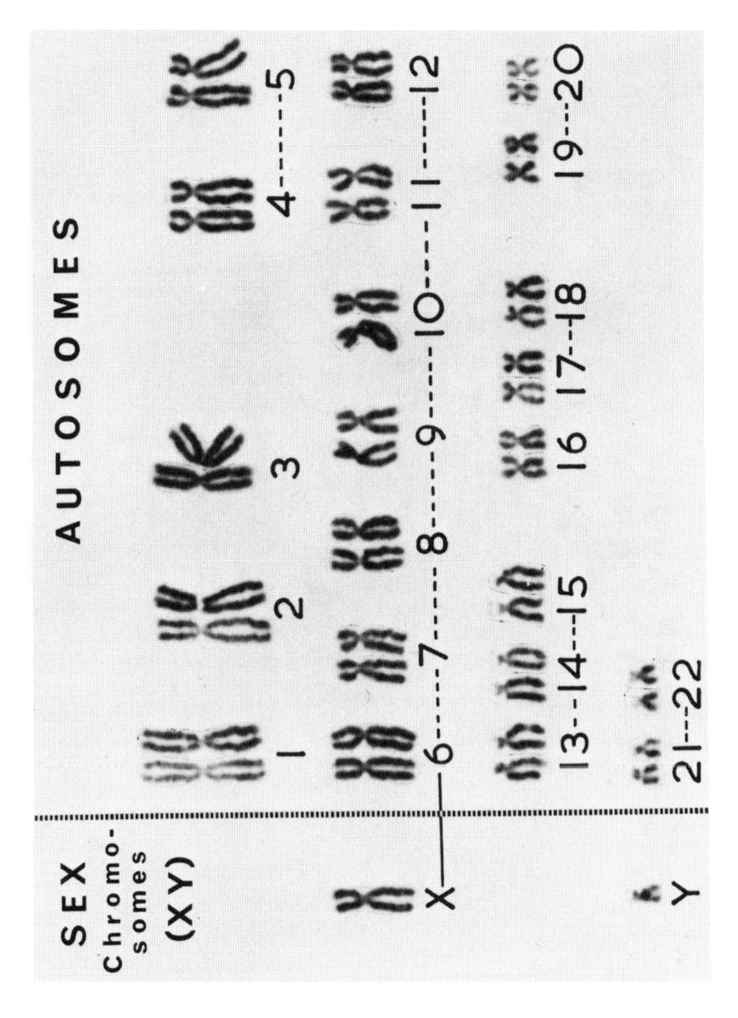

FIG. 1-11. Normal male karyotype. (By courtesy of Professor Paul Polani)

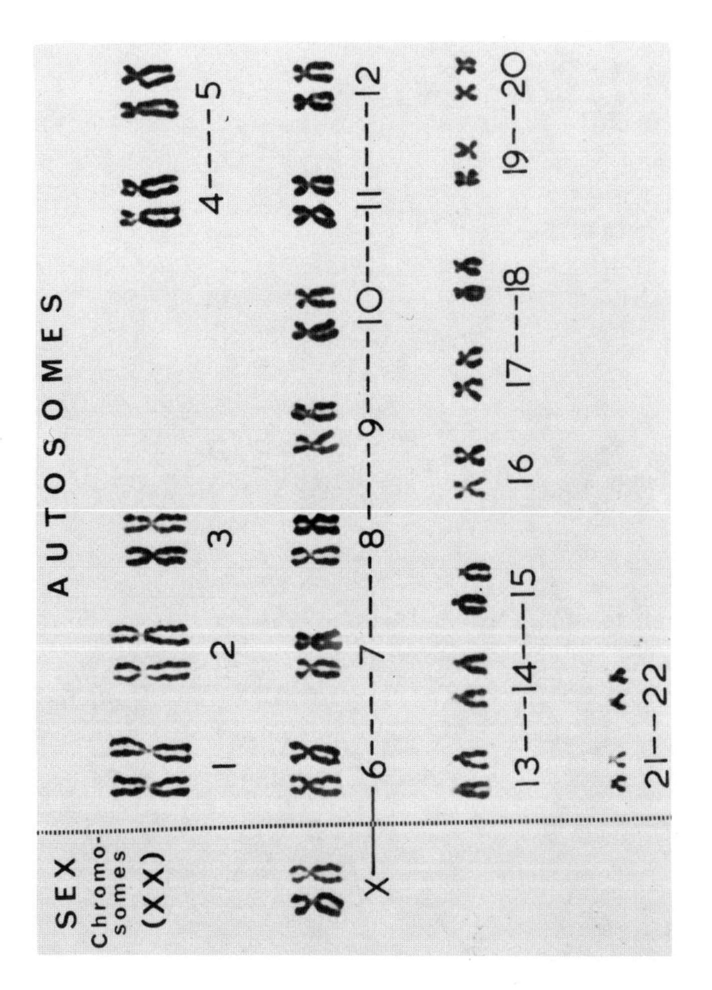

FIG. 1-12. Normal female karyotype. (By courtesy of Professor Paul Polani)

V. NUMERICAL CHANGES IN THE KARYOTYPE

Various errors may occur at mitosis or meiosis that can lead to an anomaly of chromosome number, resulting in aneuploidy or polyploidy.

1. Aneuploidy is the result of the loss or gain of chromosomes due to an error at mitosis or meiosis. Nondisjunction is the name given to this error, whereby two chromosomes or chromatids fail to go to the opposite poles of the spindle. This can occur at anaphase in mitosis, or at anaphase I or anaphase II in meiosis with differing results. An individual with an extra member of a chromosome pair is trisomic for that chromosome, the karyotype of a male with an extra chromosome 21 being 47, XY, 21+. When only one member of a chromosome pair is present there is monosomy for that chromosome, the karyotype of a female with only one chromosome 21 being 45, XX, 21-.

2. Polyploidy is another abnormality of chromosome number where there is a multiple of the normal haploid number, as for example triploidy (69, XXX or 69, XYY) and tetraploidy (92, XXXX).

The presence of several cell lines with different chromosome numbers is termed mixoploidy, although the term "chromosome mosaic" is still widely used. The karyotype of a female with two cell lines, one normal and one with a missing X chromosome, is 45, X/46, XX.

VI. STRUCTURAL CHANGES IN THE KARYOTYPE

Chromosomes may undergo structural changes during interphase, or during mitosis and meiosis. Numerical aberrations occur far more frequently than structural changes, although the actual incidence of structural changes may be higher than is reported, owing to difficulties in detection. Structural changes may be divided into deletions, inversions, insertions, and translocations.

A. Deletions

A single chromosome may sustain one, two, or more rarely, three breaks. Figure 1-13a illustrates a terminal chromosome deletion occurring after a single break. The acentric fragment is lost and the new deleted chromosome is formed after the broken end heals. An interstitial deletion will occur after two breaks, with the loss of the interstitial fragment, followed by rejoining of the two ends as in Fig. 1-13b. An interstitial deletion occurs more frequently than a terminal deletion, probably because the latter is an unstable structure that requires the formation of a new telomere. The telomere is a stable structure occurring at the ends of each chromosome and its presence prevents the broken end of a chromosome from uniting with the end of an intact chromosome. The karyotype of a female with a deletion of the short arm of chromosome 18 is 46, XX, 18p-. In the new notation (Paris Conference, 1971), the location of the break is indicated; for example, 46, XX, del(1)(q21) denotes a break at band 1q21 and loss of the long arm segment distal to it.

B. Inversions

If there are two breaks on the same arm of a chromosome and the interstitial fragment is inverted before reunion, the resultant arrangement is a paracentric inversion. These are difficult to detect except by banding, since the resultant chromosome is morphologically identical to its homologue. In addition, there is no loss of genetic material and the phenotype is usually unchanged. If the two breaks are on either side of the centromere and the interstitial fragment is inverted before reunion, the result is a pericentric inversion. These are detected far more easily, since they result in a different position of the centromere relative to the arms of the chromosome (Figs. 1-13c and 1-13d).

A pericentric inversion is denoted by inv(p+q-) or inv(p-q+). In the new notation, the points of breakage and union are indicated; for example, 46, XY, inv(1)(p22q24) denotes breaks occurring at bands 1p22 and 1q24 with the interstitial fragment being inverted before reunion, In a paracentric inversion, the break point more proximal to the centromere is always specified first as in inv(2)(q13q23).

C. Rings

A chromosome may sustain two terminal deletions, one on each arm. The broken ends unite and form a ring chromosome with the loss of two fragments. The karyotype of a male with a ring chromosome 18 is 46, XY, 18r or 46, XY, r(18)(p11q23), indicating break points at bands 18p11 and 18q23.

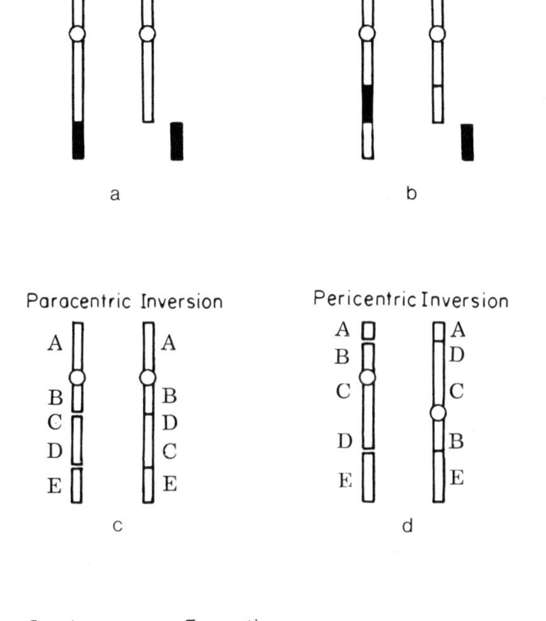

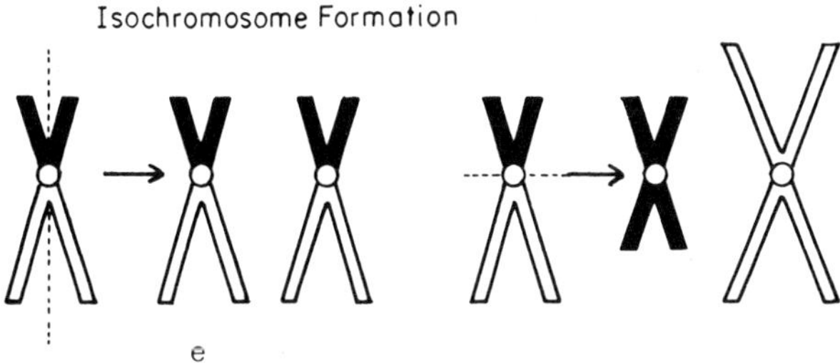

FIG. 1-13. (a) Terminal deletion; (b) interstitial deletion; (c) paracentric inversion; (d) pericentric inversion; (e) isochromosome formation showing longitudinal splitting; and (f) isochromosome formation showing transverse splitting.

D. Isochromosomes

At metaphase, the centromere in a normal chromosome splits longitudinally to form a new chromosome (Fig. 1-13e). If, however, the centromere splits in a transverse fashion (Fig. 1-13f), the resultant chromosomes are isochromosomes in which each is made up of identical arms. An isochromosome is denoted by the lower case letter "i" as for example, 46,XXqi or 46,X,i(Xq).

E. Translocations

The transfer of a chromosome segment from its normal position to another position is a translocation, and there are several different types of translocation.

1. Shifts. This is the transfer of a chromosome segment from one position to another position on the same chromosome. Shifts are extremely difficult to detect and they have only recently been reported in the literature. There is little doubt that they will be increasingly detected by banding techniques.

2. Insertions. This is the transfer of a chromosome segment into another chromosome and requires three breaks (Fig. 1-14a). These are being reported in the literature and they can be detected more easily than shifts.

3. Reciprocal Translocations. This is the exchange of segments which may be terminal or interstitial between two nonhomologous chromosomes (Fig. 1-14b). Figure 1-14b illustrates an exchange of segments after two breaks in which each chromosome has lost a terminal portion of one of its arms. In the two resultant chromosomes, T1 and T2, T1 is deficient in the length of its long arm, and T2 is duplicated in part of the length of its long arm. The total amount of genetic material is unchanged and, though the order of the bases on the DNA chain may be altered, the result is a balanced translocation and the phenotype should be normal. Using the new notation where the derivative chromosome with the lowest number is indicated first, a reciprocal translocation in a male between the short arm of chromosome 4 and the long arm of chromosome 15 would be written as 46,XY,t(4p−;15q+) or 46,XY,t(4;15)(p15;q22). Where there have been family studies and the origin of a translocated chromosome is known, this derived chromosome may be indicated by "mat" or "pat." In the previous example, a son may have inherited the deficient chromosome 4 from his father, and his karyotype would then be 46,XY,4p−pat. A balanced translocation may be transmitted in an unbalanced form, depending on the segregation or the way in which chromosomes move to the poles at meiosis. This may be illustrated in a family where the mother had a (5p−;13q+) translocation (Fig. 1-14c). At meiosis, her

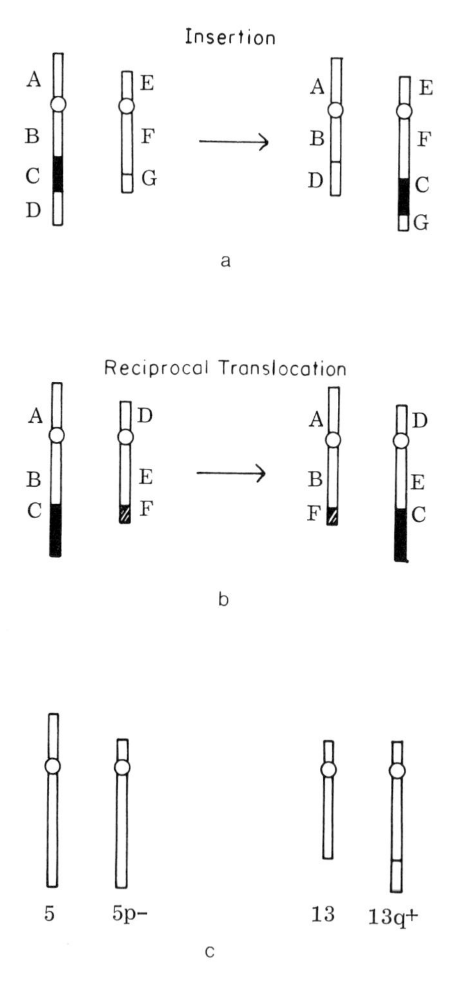

FIG. 1-14. (a) Insertion; (b) reciprocal translocation; and (c) reciprocal translocation, t(5p-;13q+).

translocated chromosomes and their homologues form a cross (Fig. 1-15a), making various segregations possible. The alternate is when chromosomes 5 and 13 go to one pole, and 5p- and 13q+ go to the opposite pole. This results after fertilization in either a normal zygote or a zygote with a balanced translocation. Another type

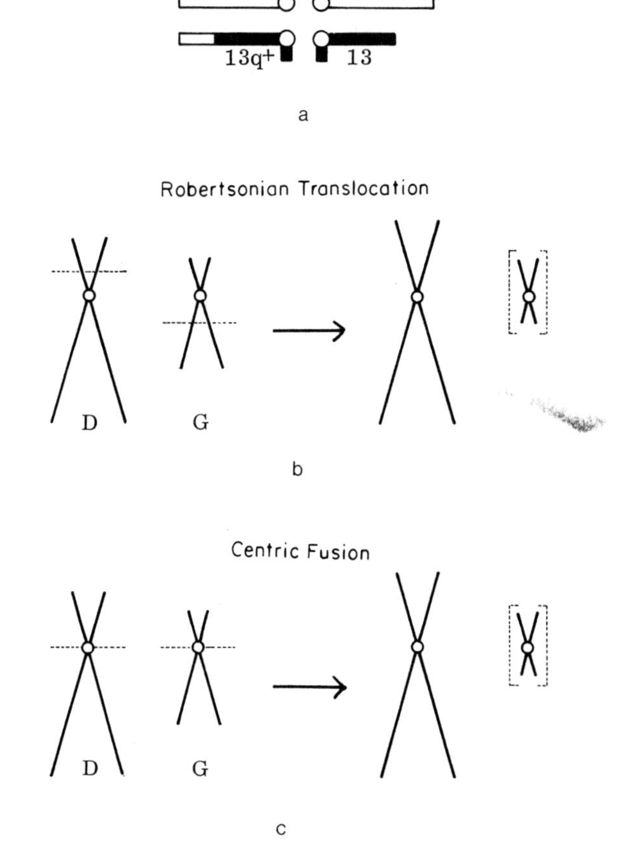

FIG. 1-15. (a) Meiotic cross with translocation chromosomes; (b) Robertsonian translocation; and (c) centric fusion.

of segregation is known as adjacent-1 in which homologous chromosomes go to opposite poles. One set of gametes will have 5 and 13q+ and result in a zygote which is trisomic for 5p, and the other set of gametes will have 5p- and 13 and a zygote which has a 5p- deletion. A third type of segregation is known as adjacent-2, where homologous chromosomes go to the same pole. In adjacent-2 segregation, one set of gametes will have 5 and 5p-, resulting in a zygote which is monosomic for 13. The other set of gametes will have 13 and 13q+, and the zygote will be monosomic for 5. In this particular family, for example, the mother had three normal children out of nine pregnancies. Two pregnancies had the 5p- deletion, two were partially trisomic for 5p, and two were abortions. It is possible that the two abortions resulted from an adjacent-2 type of segregation where there is autosomal monosomy, which is considered to be incompatible with life. The other four affected children resulted from an adjacent-1 type of segregation, and the three normal children resulted from an alternate type of segregation.

There is also the 3-1 meiotic disjunction which differs from alternate, adjacent-1 and adjacent-2, in that the chromosomes forming the cross at meiosis do not segregate in pairs but with three chromosomes proceeding to one pole of the cell and the fourth moving to the opposite pole. This type of disjunction leads to 20 derivable chromosome types and to offspring with 45 or 47 chromosomes. According to the chromosome type, there may be partial trisomy, partial monosomy, or tetrasomy of some chromosome regions. The incidence of this type of disjunction is not yet known, but four recent reports with ocular abnormalities have been included in this review in the section on aneuploid translocations.

Most of the translocations reviewed here are of the adjacent-1 type of segregation and result in a partial trisomy for a given chromosome and a partial monosomy for a small portion of another chromosome. There are also balanced translocations which are transmitted to children who are not phenotypically normal. This has been explained by the concept of "aneusomie de recombinaison" which suggests a complex rearrangement, such as a paracentric inversion in the translocated chromosome. This, in turn, gives rise to a chromosome which is morphologically indistinguishable from the translocated chromosome, but with some duplication or deficiency of the chromosome segment.

4. Robertsonian Translocations. Another type of translocation is "centric fusion" or Robertsonian translocation where two acrocentric chromosomes break on opposite sides of the centromere (Fig. 1-15b). The two long arms rejoin to form a new metacentric or submetacentric chromosome, and the two short arms form a minute centric fragment which is usually lost. In centric fusion, it is supposed that the two centromeres split (Fig. 1-15c). For example, Robertsonian translocation would be denoted as 45,XX,t(13;14)(p11;q11), with breaks on bands 13p11 and 14q11, and the segment distal to 14q11 translocated onto chromosome 13, with loss of the fragment of chromosome 14 including the centromere, and also the segment distal to 13p11. A

centric fusion would be denoted by the formula 45, XX, t(13q14q), indicating that breakage occurred near the centromere in both chromosomes 13 and 14, with loss of the short arms of both chromosomes. There are centric fusions of D/D, D/G, and G/G types, the first being one of the most common forms found in man, occurring in about one out of every thousand of the population. Depending on the type of segregation, a D/G translocation may be transmitted as a normal individual, a translocation carrier, a monosomy D or G, or a trisomy D or G. In practice, autosomal monosomy is usually incompatible with life, but there are a few reported cases of monosomy G (Chapter 8, page 139); none of these, however, was the result of a parental translocation. The trisomies 13 and 21 arising from centric fusion, are not discussed here since they should be compared with the ordinary trisomy. A few cases of DqDq translocations other than 13q13q are, however, included.

VII. THE SEX CHROMOSOMES

The normal female has two X chromosomes, the normal male has one X chromosome and a Y chromosome. The Y chromosome is purely male determining, and its action on the indifferent embryonic gonad causes it to develop as a testis. In the absence of a Y chromosome, the phenotype will be female regardless of the number of X chromosomes present. The X chromosome carries X-linked genes and female-determining factors.

All males receive their X chromosome from their mother and their Y chromosome from their father, and in turn they will transmit their X chromosome to their daughters and their Y chromosome to their sons. This is of considerable importance in the transmission of X-linked diseases such as ocular albinism, the oculo-cerebro-renal syndrome of Lowe, choroideremia, X-linked retinitis pigmentosa, to name but a few. The whole topic of X-linked ocular diseases can only be mentioned briefly here, but it is of great interest to ophthalmologists as the detection of the heterozygous carrier females is gradually becoming possible.

1. Sex Chromatin

If female somatic cells are stained in early mitosis, one X chromosome stains differently from the other X chromosome and from the autosomes. This X chromosome is said to be heterochromatic while the autosomes and the other X chromosome are euchromatic. In somatic cells, the XY male and the XX female have one euchromatic X chromosome in each nucleus. The heterochromatic X chromosome constitutes the Barr body or sex chromatin, which is a chromatin mass present in the interphase nuclei of normal females but which is absent in normal males. The Barr body is present in nearly all female somatic cells and, for the diagnosis of chromosomal sex, is usually studied in the cells of the buccal mucosa. The number of Barr bodies in any diploid cell is always one less than the number of X chromosomes present. Thus, there is one Barr body with XX, XXY, and XXYY; two Barr bodies with XXX, XXXY, and XXXYY (Fig. 1-16). The size of the Barr body is also related to the structure of the heterochromatic X chromosome; an abnormally small Barr body is associated with a deficient X chromosome such as Xp-, Xq-, Xr or Xpi, while a large Barr body is found with a duplicated X chromosome such as Xqi.

2. The Lyon Hypothesis

The normal male has a single X chromosome and only a single dose of each X-linked gene, while the normal female with two X chromosomes seems to have a double dose of each X-linked gene.

Obviously there has to be some kind of dosage compensation mechanism to allow for the same phenotypic expression of an X-linked gene in the male and in the female. This is explained by the Lyon hypothesis which states that very early in embryogenesis, one of the X chromosomes in the female becomes genetically inactive and is heterochromatic. The heterochromatic X chromosome is inactivated at random; it may be maternal or paternal in origin in the same organism, and once inactivation has occurred it remains fixed throughout the ensuing development of each cell line. If there are more than two X chromosomes present, all but one is inactivated.

A consequence of the Lyon hypothesis is that all normal females are effectively mosaics, with one cell line derived from a paternally active X chromosome, and the other cell line derived from a maternally active X chromosome. The female heterozygous for an abnormal gene on the X chromosome shows variable expressivity for this gene, since the inactivation of one of her X chromosomes is entirely at random, the expressivity of this abnormal gene will vary according to the proportion of active X chromosomes carrying this gene. This is found to be the case in female heterozygotes with, for example, ocular albinism, choroideremia, or X-linked retinitis pigmentosa.

The case of 45,X females (Turner syndrome) is also very interesting. They have only one active X chromosome and are chromatin

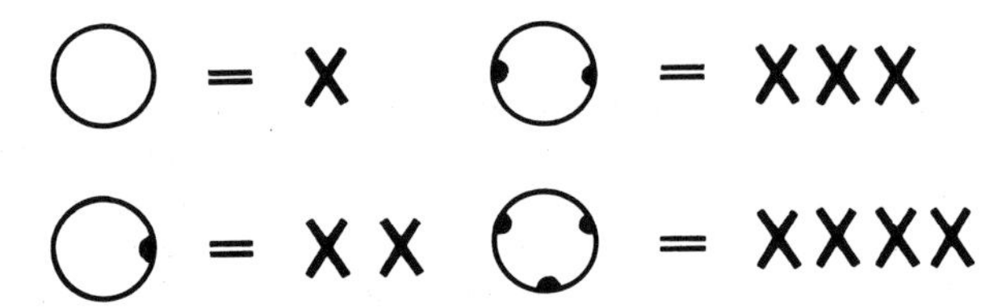

FIG. 1-16. Number of Barr bodies in association with X chromosomes. Each X chromosome in excess of one forms a sex-chromatin mass. (By courtesy of Professor Paul Polani)

negative, that is to say they have no Barr body. They are sterile and have ovarian dysgenesis, their gonads are streaks with no germ cells. The gonads of 45, X fetuses, however, are relatively normal up to about the third month of gestation. The germ cells degenerate after this time, there are far fewer than in the normal 46, XX ovaries and they disappear altogether by the time of puberty. It seems therefore that two X chromosomes are necessary to maintain normal ovarian development, so that the inactivated X chromosome plays some part in this development. It has been suggested that only part of the X chromosome is inactivated, the active part being the short arm, and that Turner syndrome is due to monosomy of loci on the short arm of the X chromosome which are homologous with loci on the long arm of the Y chromosome.

In females with a structurally abnormal X chromosome, Xp-, Xq-, Xr or Xqi, there is preferential inactivation of the abnormal X chromosome. There are however phenotype differences between these females and 45, X females. These occur because both X chromosomes are active in the early embryo (up to about 16 days), and because the short arm of the X remains active in both chromosomes (except in cases with 46, XXp-).

VIII. SUGGESTED READINGS

Carter, C. O., An ABC of Medical Genetics, London, The Lancet, Ltd., 1969.

Ford, E. H. R., Human Chromosomes, New York and London, Academic Press, Inc., 1971.

Hamerton, J. L., Human Cytogenetics, Vols. I and II, New York and London, Academic Press, Inc., 1971.

McKusick, V.A. Human Genetics, Englewood Cliffs, New Jersey, Prentice-Hall, Inc., 1969.

Ohno, S. Sex Chromosomes and Sex-linked Genes, Berlin, Springer-Verlag, 1967.

Stern, C. Principles of Human Genetics, San Francisco, W. H. Freeman & Co., 1973.

Whittinghill, M. Human Genetics and its Foundations, New York, Reinhold Publishing Corporation, 1965.

CHAPTER 2

INCIDENCE OF CHROMOSOMAL ABNORMALITIES

I. SPONTANEOUS ABORTIONS

The incidence of chromosomal abnormalities in spontaneous abortions is much higher than in liveborn infants. There have been several studies published on this topic, of which a very comprehensive review is that of Larson and Titus (6). A total of 825 karyotypes described in the literature was analysed with these results: (a) the number of major chromosomal abnormalities was 194 (23. 5%), and (b) the number of translocations was six (0. 72%). There is also a detailed description of the pathology of eyes of embryos by Howard et al. (4). The eyes from 19 embryos were studied, the karyotypes included monosomy X, triploidy, and trisomy B, C, D, and G. Some findings corresponded to a retardation of development, while others represented a dysgenesis of ocular structures. Abnormalities included cyclopia, microphthalmos, cataract, staphyloma of the optic nerve, retinal dysplasia, and subluxation of the lens.

II. PERINATAL DEATHS

Machin (9) examined 726 infants and found 28 with chromosomal abnormalities, accounting for 9% of macerated stillbirths, 4% of fresh stillbirths, and 6% of early neonatal deaths. In this study, the aneuploidies occurred most frequently. There was, however, one case of Gp- and one of a 15p+ chromosome.

III. NEWBORN INFANTS

Several large series of newborn infants have been studied, and as in the series on spontaneous abortions, the abnormalities are chiefly aneuploidies. The results of a selected number of surveys are compared in Table 2-1 where the major chromosomal abnormalities include aneuploidies, translocations, and deletions. The incidence of deletions is much smaller than that of translocations and is given wherever it is known in the appropriate sections on deletion syndromes (see pages 34, 44, and 149).

The translocations listed are of the reciprocal type, and of these the Robertsonian D/D is one of the most common forms found in about one in a thousand of the population. It is estimated that the incidence of unbalanced translocations is about 0.04%, while the average incidence of reciprocal translocations is 0.181%. An incidence of 0.5% for a major chromosomal abnormality was found by Lubs and Ruddle (7), who calculated that with this incidence, 20,000 infants are born every year in the United States with a major chromosomal abnormality. Assuming an incidence of 0.181%, there are about 7,000 infants born every year in the United States with a reciprocal translocation. As can be seen in the following chapters, nearly all children with an unbalanced translocation (a chromosome duplication) have some degree of mental subnormality and various congenital abnormalities. Even if these children inherit a translocation in a balanced form and have no phenotypic abnormality, there is an increased risk they will give birth to affected offspring. The risk of mothers with D/G translocation giving birth to a child with trisomy 21, is one in six, but the risk is less if the fathers are translocation carriers. This is due to the decreased motility of sperm carrying abnormal chromosomes. According to Gerald and Walzer (2), four out of five of the reciprocal translocations they found were inherited, and it would seem that only a small proportion of reciprocal translocations are new mutations.

TABLE 2-1

Incidence of Chromosomal Abnormalities

Survey	Total births	Number of major chromosomal abnormalities		Number of translocations	
Sergovich et al. (10)	2159	10	(4. 8/1000 births)	1	(0. 46/1000 births)
Lubs and Ruddle (8)	4400	22	(5. 0/1000 births)	6	(1. 36/1000 births)
Gerald and Walzer (2)	3543	18	(5. 0/1000 births)	5	(1. 41/1000 births)
Friedrich and Nielsen (1)	5049	43	(8. 5/1000 births)	16	(3. 17/1000 births)
Jacobs et al. (5)	11680	78	(6. 7/1000 births)	22	(1. 88/1000 births)
Hamerton et al. (3)	6809	22	(3. 2/1000 births)	11	(1. 62/1000 births)
Totals	33640	193	(5. 7/1000 births)	61	(1. 81/1000 births

REFERENCES

1. Friedrich, U. and Nielsen, J. Chromosome studies in 5049 consecutive newborn children. Clin. Genet. 4: 333-343 (1973).

2. Gerald, P. S. and Walzer, S. Chromosome studies of normal newborn infants. In Human Population Cytogenetics, P. A. Jacobs, W. H. Price, and P. Law (Eds.), London, Edinburgh University Press, 1970, pp. 143-151.

3. Hamerton, J. L., Ray, M., Abbott, J., Williamson, C., and Ducasse, G. C. Chromosome studies in a neonatal population. Can. Med. Assoc. J. 106: 776-779 (1972).

4. Howard, R. O., Boué, J., Deluchat, Ch., Albert, D. M., and Lahav, M. The eyes of embryos with chromosome abnormalities. Am. J. Ophthalmol. 78: 167-188 (1974).

5. Jacobs, P. A., Melville, M., Ratcliffe, S., Keay, A. J., and Syme, J. A cytogenetic survey of 11,680 newborn infants. Ann. Hum. Genet. 37: 369-376 (1973 and 1974).

6. Larson, S. L. and Titus, J. L. Chromosomes and abortions. Mayo Clin. Proc. 45: 60-72 (1970).

7. Lubs, H. A. and Ruddle, F. H. Chromosome abnormalities in the human population: Estimation of rates based on New Haven newborn study. Science 169: 495-497 (1970).

8. Lubs, H. A. and Ruddle, F. H. Applications of quantitative karyotypy to chromosome variation in 4400 consecutive newborns. In Human Population Cytogenetics, P. A. Jacobs, W. H. Price, and P. Law (Eds.), London, Edinburgh University Press, 1970, pp. 119-142.

9. Machin, G. A. Chromosomal abnormality and perinatal death. Lancet 1: 549-551 (1974).

10. Sergovich, F., Valentine, G. H., Chen, A. T. L., Kinch, R. A. H., and Smout, M. S. Chromosome aberrations in 2159 consecutive newborn babies. N. Engl. J. Med. 280: 851-855 (1969).

CHAPTER 3

DEFICIENCIES OF GROUP B CHROMOSOMES

I. INTRODUCTION

There are several deficiency syndromes such as 4p-, 5p- and the deficiencies of chromosomes 13 and 18, which are well documented and for which there are many associated ocular abnormalities. Each of these will be described in its own chromosome group with tables listing the ocular abnormalities as reported in the literature. Deficiencies of group A chromosomes with associated ocular abnormalities have not been found in the literature, although a few duplications involving group A chromosomes, resulting in partial trisomies, have been reported and these will be discussed in their appropriate section.

II. 4p- OR WOLF-HIRSCHHORN SYNDROME

The first reference in the literature to what has become known as the Wolf-Hirschhorn syndrome, is that of Hirschhorn and Cooper (38), who described a child with defects of midline fusion, a coloboma of the right iris, and an apparent deletion of the short arm of chromosome 4. Hirschhorn et al., (39) describe what one may assume to be the same child, in that the titles of the two papers are the same except for one adjective, with a child presenting the same symptoms. Two other early references are those of Wolf et al., (118, 120), one written in German and the other in English but both presumably describing the same child with hypertelorism, prominent glabella, left-sided cataract, coloboma and ptosis, and ectopia of the right pupil. In the German article, the child also had stenosis of the left nasolacrimal duct and horizontal nystagmus. This article is also the first in which the deficiency of the short arm of this chromosome was ascribed by autoradiography to No. 4; other previous references (21, 38, 40, 102) had referred to the deleted short arm of the chromosome belonging to group B. All references subsequent to that of Wolf et al., (120) cite chromosome 4 except for Ricci et al., (92) where the chromosome is of group B.

A. Incidence

References to the incidence of the 4p- syndrome are difficult to find but according to Ford (23) they are clearly very low, while Polani (87) estimates that perhaps one-third of the Bp- may be 4p- syndromes. Hamerton (35) suggests that children with a 4p- deletion may be ascertained as possible 13-trisomics, or that chromosome 5 may be more susceptible to breakage than chromosome 4, or that a deletion of chromosome 4 may be more lethal in fetal life, although he adds that this last assumption is not borne out by any evidence from chromosome studies on spontaneous abortions.

B. Phenotype

The characteristic defects of the 4p- syndrome include microcephaly, profound psychomotor retardation, defects of midline fusion, a broad root of the nose, asymmetry of the face, prominent glabella, ocular abnormalities, electroencephalogram (EEG) abnormalities, and hypospadias in boys. Other abnormalities encountered include micrognathia, midline scalp defects, low ears with little cartilage, sacral and other skin dimples, flexion deformities of the fingers, clinodactyly, transverse palmar creases, and malformed big toes (116). It is interesting to note that these abnormalities are more severe than those of the 5p- syndrome.

C. Ocular Abnormalities

The most common findings in 43 cases of 4p- (Table 3-1) are hypertelorism (30 cases, 70%), antimongoloid slant (11 cases, 26%) epicanthus (13 cases, 30%), convergent strabismus (11 cases, 26%), divergent strabismus (7 cases, 16%), and coloboma of the iris (13 cases, 30%) (Fig. 3-1). Ptosis has been reported (33, 85, 118, 120), while van Kempen and Jongbloet (48) describe their patient as having a deep fold under both eyes, bilateral lid lag and ectropion of the

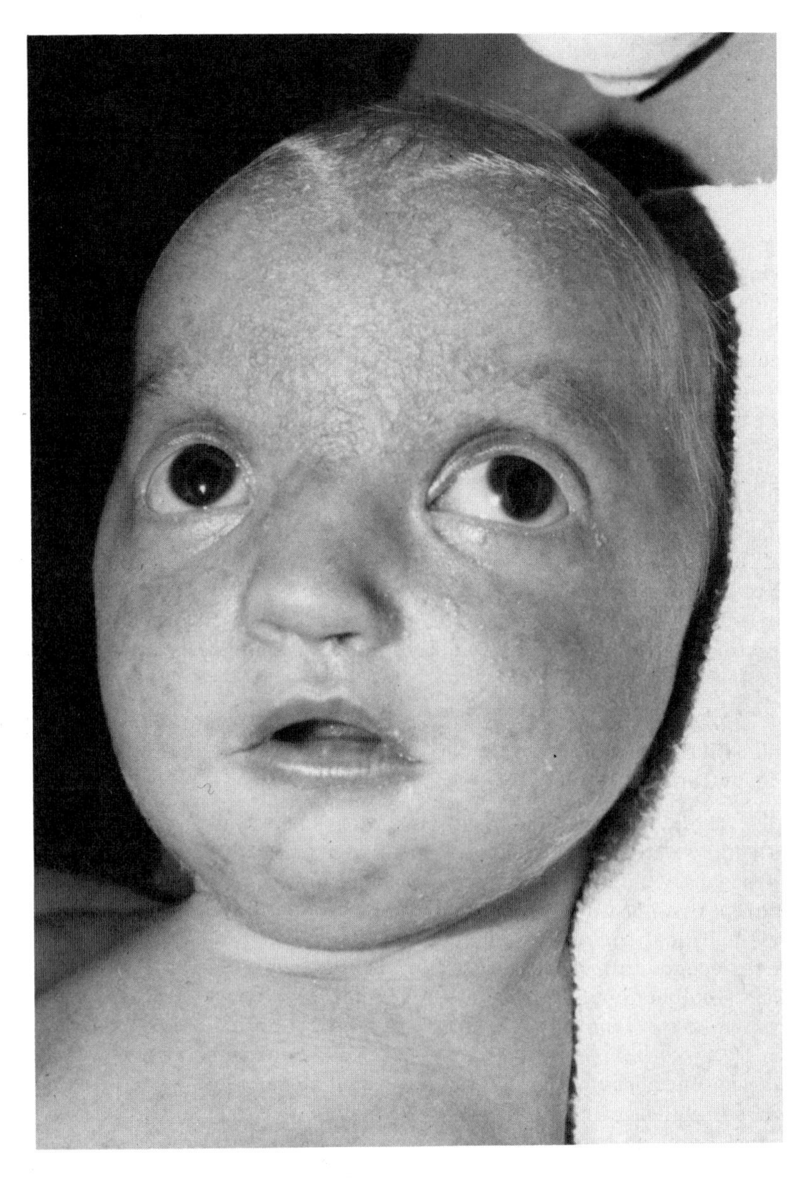

FIG. 3-1. 46, XY, 4p- showing bilateral iris colobomas and divergent strabismus. (By permission of Mr. L. J. Butler)

upper lids, as well as other ocular abnormalities. François et al., (24) described two of their cases as having a syndrome of palpebral retraction with a retracted lid and they attribute this to hypertonicity of all skin muscles.

Exophthalmos has been reported by many investigators (33, 48, 53, 54, 85, 108, 115, 116). A fixed pupil has been described (115), as has an ectopic pupil (48, 115, 118). These pupillary abnormalities are considered by François et al., (24) as a minor form of the anterior chamber cleavage syndrome. Wolf et al., (120) describe stenosis of the nasolacrimal duct, as do François et al., (24) in two of their patients. Abnormalities of the cornea are rare: bilateral corneal opacities were found in case 2 of Taylor et al., (111), and the first case described by François et al., (24) had sclerocorneae, although much thinner in the left eye than in the right, iris colobomas, no anterior chamber, and bilateral anterior synechiae. Another rare finding is nystagmus (116, 118, 120) and possible nystagmus in three out of four cases (33). Abnormalities of the iris such as colobomas (see above), speckled irides (1, 54) and Brushfield spots (24, 116) are frequently found. Cataracts have been reported in only three cases (45, 85, 118, 120). In addition to a cataract, Pfeiffer's patient (85) had what is best described as a sector-shaped posterior persistent tunica vasculosa lentis and pigment cell proliferation.

A few cases with a 4r (11, 29, 36), Br (16) or Bq- (62, 78) chromosome have been included as the clinical appearance is similar to that described in cases of 4p-.

There is also the atypical case of Jalbert et al., (45) who described a child with the Hallermann-Streiff syndrome (oculomandibulofacial dyscephaly with hypotrichosis). The child was born of consanguineous parents, he had dwarfism, a bird-like face, bilateral microphthalmos with corneal opacities, a leucomatous cornea of reduced size in the right eye, congenital cataract with bilateral remnants of pupillary membranes, and a coloboma of the left iris. The karyotype showed a Bp- chromosome which was presumed to be a 4p-, but the child died before autoradiography could be performed. This rare 4p- syndrome is associated with major congenital abnormalities, many of which seem to be related to a developmental defect of midline fusion. It may be that the ophthalmic manifestations, such as hypertelorism, slanted palpebral fissures and strabismus, are related to this defect of midline fusion. Iris colobomas, however, are the result of a defect in the closure of the fetal fissure of the optic cup, and they occur in about 30% of cases of 4p-. Iris colobomas are associated with other chromosomal abnormalities, notably trisomy 13, but it is interesting to note that colobomas that are unrelated to chromosomal abnormalities are not infrequently associated with skeletal abnormalities and craniofacial dystrophies (19). It would seem, therefore, that in this syndrome the ocular abnormalities are the result of a defect of midline fusion; iris colobomas are the exception and may be an expression of faulty development occurring at the time when the fetal fissue is closing.

TABLE 3-1

Group B Chromosomes: 4p-

Reference	Ptosis	Antimongoloid slant	Hypertelorism	Epicanthus	Microphthalmos	Exophthalmos	Strabismus	Ectopic pupil	Iris coloboma	Cataract	Chromosome abnormality	Remarks
Hirschhorn and Cooper (38) 1961									(+)		Bp-	
Sidbury et al. (102) 1964		+	+						(+)		Bp-	
Dyggve and Mikkelsen (21) 1965			+	+							Bp-	
Hirschhorn et al. (39) 1965									(+)		Bp-	
Wolf et al. (118, 120) 1965a, b	(+)		+					(+)	(+)	(+)	4p-	Stenosis of the left nasolacrimal duct; horizontal nystagmus

Leao et al. (54) 1966					(+)	+					4p-	Hypoplasia of left orbit; partial 4p-
Miller et al. (67) 1966	Case 1			+			Con				4p-	
	Case 2		+	+	+		Con				4p-	
Ricci et al. (92) 1966	Case 1		+	+	+						4p-	
	Case 2		+	+	(+)						Bp-	
Leao et al. (53) 1967				+		+	Con				4p-	"Irides coarse blue and marked with white" Partial 4p-
van Kempen and Jongbloet (48) 1967				+		+		+	+		4p-	Bilateral lid lag; ectropion of the upper lids
Ockey et al. (78) 1967											4q-	Slight mongoloid slant
Jalbert et al. (45) 1968					+				(+)	+	4p-	Hallermann-Streiff syndrome: Left, sclerocornea; right, microcornea; pupillary membrane
Pfeiffer et al. (85) 1968	Case 1	+		+						+	4p-	Partial tunica vasculosa lentis; "pigment cell proliferation"
	Case 2	+		+		+					4p-	

TABLE 3-1 (continued)

Reference		Ptosis	Antimongoloid slant	Hypertelorism	Epicanthus	Microphthalmos	Exophthalmos	Strabismus	Ectopic pupil	Iris coloboma	Cataract	Chromosome abnormality	Remarks
Carter et al. (11) 1969										(+)		4r	Marked proptosis; clinical appearance similar to 4p−
Dallaire (16) 1969						(+)						Br	
Hecht (36) 1969					+							4r	
Šubrt et al. (108) 1969				+			+	Con				4p−	
Warburton et al. (114) 1969	Case 28			+	+			+		0		4p−	Oblique palpebral fissures
	Case 29			+	0			+		+		4p−	Oblique palpebral fissures
Arias et al. (3) 1970	Case 1			+	+					0		4p−	One case with speckled iris
	Case 2			+	+					0		4p−	

Miller et al. (68) 1970	Case 1		(+)	+	0		Div		4p-	
	Case 2		+	+	0		Div		4p-	
	Case 3		+	+	0		Div		4p	
Miller et al. (68) 1970	Case 4		0	+	+		Div		4p-	
	Case 5		+	+	+		Div		4p-	
Passarge et al. (81) 1970			+	+			Div		4p-	"Mild divergent strabismus"
Taylor et al. (111) 1970	Case 1			+				+	4p-	Marked ocular hypertelorism
	Case 2			+					4p-	Extremely marked hypertelorism; bilateral corneal opacities
Wilcock et al. (115) 1970						+		+	4p-	RE: fixed pupil. LE: ectopic pupil
Wilson et al. (116) 1970		+	+	+		+	+		4q+	Nystagmus; Brushfield spots
Citoler et al. (14) 1971				+				+	4p-	Megalocornea
Guthrie et al. (34) 1971	Case 1		+	+	+		+	0	4p-	
	Case 2			+	+	0	?	0	4p-	Cases 2, 3 and 4 had ? nystagmus

TABLE 3-1 (continued)

Reference	Ptosis	Antimongoloid slant	Hypertelorism	Epicanthus	Microphthalmos	Exophthalmos	Strabismus	Ectopic pupil	Iris coloboma	Cataract	Chromosome abnormality	Remarks
Guthrie et al. (34) 1971 Case 3			0	0		+	+		0		4p−	
Case 4	+		+	0		+	+		+		4p−	
Lindenbaum and Butler (162) 1971											Bq−	Fundi showed traces of black pigment
Moyer (72) 1971							+		(+)		4p−	
François et al. (24) 1972 Case 1									+		4p−	No anterior chamber; solid anterior synechia; bilateral stenosis of lacrimal ducts
Case 2									+		4p−	10 D myopia; Brushfield spots
Case 3											4p−	Posterior bilateral embryotoxon; stenosis of lacrimal ducts
Case 4											4p−	Retraction of the lids

	Case 5					Div	4p-	Bilateral papillary colobomas
Fryns et al. (25) 1973	Case 1		+	+	+		4p-	
	Case 2			+			4p-	
Ginsberg and Soukup (29) 1974		+	+				4r	Abnormalities of cornea, lens, vitreous, uvea and retina
Judge et al. (46) 1974					+		4p-	Proptosis; mongoloid slant

III. 5p- OR CRI-DU-CHAT (CAT CRY) SYNDROME

The first account of 5p- or cri-du-chat (cat cry) syndrome was published by Lejeune et al., (59) when they described three cases with a partial deletion of the short arm of chromosome 5.

1. Incidence

This is a rare syndrome with an estimated incidence of between one in 50,000 and one in 100,000 births or less (87), and consequently some of the cases published since Lejeune's original paper have been described by several groups of authors. Although this review includes 130 observations, the actual number of cases is more likely to be about 110, due to the duplication of data. An attempt has been made, in some measure, to report the incidences of the most common ocular abnormalities found in this syndrome, and which have been described in the literature. It is obvious that these figures cannot be strictly accurate due to the duplication of data, and also to the possibility that in a few cases a minor ocular abnormality may have been present but not described.

2. Phenotype

The most consistent finding in children with the cri-du-chat syndrome is microcephaly, reported by Taylor (110) in all the 17 cases she reviewed. Other abnormalities include failure to thrive, mental and developmental retardation, the abnormal cry in infancy, micrognathia, abnormally shaped ears, skeletal abnormalities, and numerous ocular abnormalities. These are all mild congenital malformations in association with profound mental retardation, so that a number of patients have survived childhood. Breg et al., (8) studied 13 older patients, all of whom are in institutions for the mentally subnormal. Adolescents and adults with this syndrome have also been studied by other authors (41, 76, 105). The abnormal cry of these infants has been compared to the mewing of a distressed kitten, and is due to an immature larynx. The voice patterns of infants with this syndrome have been analysed by Legros and Van Michel (55). Another distinctive feature of this syndrome is the "moonlike" face which, when associated with low-set ears, hypertelorism, epicanthus and often with divergent strabismus, gives these children their very characteristic appearance, and when associated with mental retardation often results in their detection (Fig. 3-2).

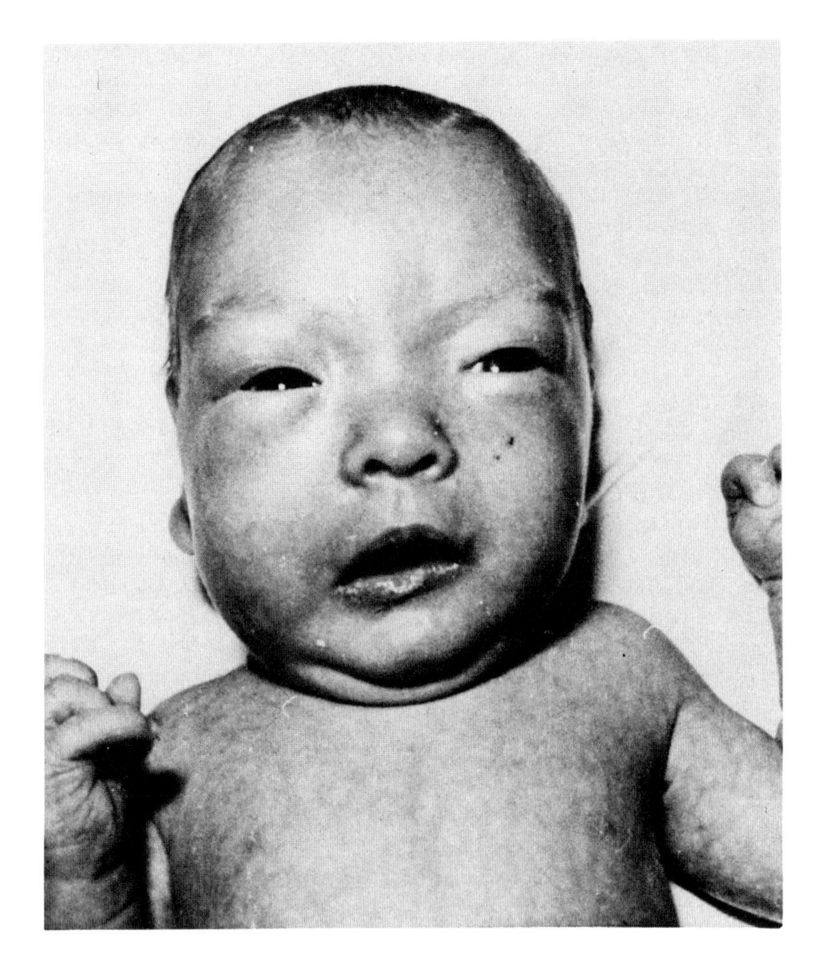

FIG. 3-2. A case of 5p- showing "moon face" and hypertelorism. (By permission of Dr. Renata Lax and Dr. Michael Ridler)

3. <u>Ocular Abnormalities</u>

The most common ocular abnormalities found in the literature are listed in Table 3-2. Only one author, Howard (41), has made a detailed study of ocular abnormalities in this syndrome, and from his findings it is very possible that other cases described in the literature may have had certain ocular abnormalities which were not recorded.

The usual minor ocular abnormalities associated with other chromosomal anomalies are also present in this syndrome: epicanthus (75%), hypertelorism (70%), and antimongoloid slant (55%). Although microcephaly is not an ocular abnormality it has been included in Table 3-2 for the purpose of comparison with the findings of other authors and because of its consistent finding, whereas hypotelorism is not found. Divergent strabismus, also, is far more common than convergent strabismus. Nystagmus is a rare finding reported by Colover et al., (15) in a boy who could read fluently and was therefore far from being typical of this syndrome. As a rule, the iris does not seem to be affected and no reports were found of colobomas or abnormalities of the anterior chamber. The lenses are not often affected; Lejeune and coworkers (60) reported opalescent lenses without opacities in one of their two cases, and Grotsky et al., (31) described a patient with cataracts. They considered, however, that the occurrence of cataracts and the cri-du-chat syndrome in the same patient may have been coincidental.

In the cri-du-chat syndrome the retina may be affected and various abnormalities have been described. The first to describe an anomaly of the fundus with pale papillae and a "pepper-and-salt" retina were de Grouchy and coworkers (33). François et al., (24) also described a pepper-and-salt anomaly of the macular region. Engel et al., (22) reported a patient with hypoplasia of the retina, choroid and optic nerves, and also a deficiency in choroidal and retinal epithelial pigmentation. A diffuse retinal atrophy was described by Laurent et al., (52) in one of three affected brothers out of a sibship of seven. Oval papillae were seen by Lichtenberger and Morineaud (61) in their patient, and incomplete bilateral demyelination of the optic nerves was reported by Noël et al., (77) in one of their cases. In his very detailed study of seven older patients, Howard (41) found optic atrophy in two, absent foveal reflexes in one, and tortuous retinal arteries and veins in six. Howard also found that the pupils were supersensitive to 2.5% methacholine in five of his cases and that there was a deficiency of tears in all seven.

4. Cytogenetics

Miller et al., (69) studies the DNA replication patterns associated with the long arm length of chromosomes 4 and 5 and established that No. 4 is late replicating and No. 5 is early replicating. One of the interesting aspects of the cri-du-chat syndrome is the balanced translocation found in one of the parents of some of the affected children (see Table 3-2). While most of the deletions of chromosome 5 arise de novo, there is a group of cases resulting from a parental balanced translocation. This occurs in about 10% of the cases in this review, which is the figure quoted by other authors (24). The parental balanced translocation is usually of the 5/D type, although 5/G and 4/5 translocations have been reported (see Table 3-2). In such families, the children may be normal or have a 5p- chromosome, or they may be partially trisomic for the short arm of No. 5. Two such families have been studied (52, 58). Noel et al., (77) described a family with a balanced translocation t(5p-;Gp+) with one normal sib, two sibs with 5p-, and the fourth sib with a Gp+ chromosome.

For comparison, a number of so-called cri-du-chat observations have been included in Table 3-2, although they are not due to a 5p- chromosome. They have various other karyotypes which may be either normal (64), or a pericentric inversion 5p-q+ with a ring C chromosome and a 15q+ chromosome (12). Many investigators have also reported chromosomal mosaicism (2, 74, 82, 93, 95, 107, 112, 121). Some of these atypical cri-du-chat cases show features uncommon to the syndrome as, for instance, that of Petit et al., (83), whose patient had a complex karyotype including 48,XY, G+F+; 49,XY, G+F+C+; 51,XY, G+G+F+C+D+; all clones were 5p-, some cells were t(5p-;Gp+), and clinically, this patient had Brushfield spots and a mongoloid slant of the palpebral fissures.

However, these are not typical cases of the syndrome which is characterized by the phenotype given above and by the partial or total deletion of the short arm of chromosome 5. From the ocular point of view the abnormalities are not quite so distinctive since they are mainly concerned with minor abnormalities of the orbits, hypertelorism or telecanthus, epicanthus and strabismus. The retina does not have a characteristic appearance as, for instance, in trisomy 13 with its retinal dysplasia, retinal folds and rosettes, nor is there a high incidence of a specific abnormality such as retinoblastoma as in the 13q- deletions. Nevertheless, the children with the cri-du-chat syndrome have a sufficiently characteristic facial appearance which combined with their mental retardation, make them instantly recognizable. It is also a fact that 10% of cases arise out of a parental translocation and it is of great importance in genetic counselling that this knowledge could prevent the occurrence of a second affected child in the same family.

It is interesting to compare the ocular abnormalities which the 4p- and 5p- syndromes have in common, and also in which respects they differ. Similar comparisons have been made (70, 117). The features common to both syndromes are hypertelorism, epicanthus, antimongoloid slant, and divergent strabismus, although the last three signs are almost twice as common in 5p- as in 4p-. The features which distinguish the two syndromes are exophthalmos, ptosis and iris coloboma, which occur in 4p- but not in 5p-.

Since the advent of autoradiography, however, all authors are unanimous in pointing out the difficulty in making a diagnosis on clinical grounds alone, and insist on autoradiographic studies in order to distinguish between the two syndromes.

TABLE 3-2

Group B Chromosomes: 5p-

Reference	Microcephaly	Epicanthus	Hypertelorism	Antimongoloid slant	Mongoloid slant	Strabismus	Optic atrophy	Chromosome abnormality	Familial translocation	Remarks
Lejeune et al. (59) 1963	3/3	3/3	3/3	3/3		1/3		5p-		Case 2: Astigmatism and intermittent divergent strabismus
DeAlmeida et al. (17) 1964		+	+	+				5p-		
Dumars et al. (20) 1964		+	+					5p-		
German et al. (27) 1964 Case 1		+	+					5p-		Described by Lejeune et al. (59), Case 1
Case 2		+	+					5p-		Oblique palpebral fissures; described by MacIntyre et al. (65)
de Grouchy et al. (33) 1964		+	+			Div		5p-		Pale papillae; "pepper-and-salt" retinae

TABLE 3-2 (continued)

Reference	Microcephaly	Epicanthus	Hypertelorism	Antimongoloid slant	Mongoloid slant	Strabismus	Optic atrophy	Chromosome abnormality	Familial translocation	Remarks
Lejeune et al. (56) 1964a	+	+	+	+		+		5p−		5th published case of 5p−
Lejeune et al. (60) 1964c Case 1		+	+	+				5p−		5th and 6th published cases of 5p−
Case 2		+	+	+		Div		5p−		Intermittent divergent strabismus; astigmatism
MacIntyre et al. (65) 1964	+	+	+	+				5p−		This case also described by Miller et al. (69) Case B
Punnett et al. (88) 1964		+	+			Div		5p−		
Berg et al. (5) 1965	+	0	0	0		0		5p−		
Bergman et al. (6) 1965			+	+		+		5p−		

Bettecken et al. (7) 1965		+	+	+	Div	5p-		
Hijmans and Shearin (37) 1965		+	+	+	Div	5p-		Facial asymmetry
Genest et al. (26) 1965	+	+	+	+	Div	5p-		
Giorgi et al. (28) 1965			+	+		5p-		Normal fundus
de Grouchy et al. (32) 1965	+	+	+	+		5p-	5/G	Maternal translocation
Holboth and Mikkelsen (40) 1965		+		+	Div	5p-		Slight antimongoloid slant
Hustinx and Wijffels (42) 1965	+	+	+	+	Alt	5p-		LE: amblyopia
Koulischer and Quersin (49) 1965		+				5p-	5/15	Paternal translocation
Lejeune et al. (57) 1965 Case 1						5p-	5/13	Maternal translocation. Presumed 5p-, died in 1945
Case 2						5p-	5/13	Described previously as Case 1, Lejeune et al. (59)

TABLE 3-2 (continued)

Reference	Microcephaly	Epicanthus	Hypertelorism	Antimongoloid slant	Mongoloid slant	Strabismus	Optic atrophy	Chromosome abnormality	Familial translocation	Remarks
McCracken and Gordon (63) 1965		+	+	+		Div		5p-		
Pfeiffer and Simon (84) 1965	+	+	+					5p-		
Reinwein and Wolf (90) 1965	+		+	+		Div		5p-	5/D	Parents had a normal karyotype
Rohde and Tompkins (94) 1965	+	+	+	+				5r		
Ricci et al. (91) 1965 Case 1		+	+	+				5p-		
Case 2			+	+				5p-		

Engel et al. (22) 1966				+			5p-	Hypoplasia of the retina, choroid and optic nerves; deficiency in choroidal and retinal epithelial pigmentation; replacement of one G chromosome by a minute centric fragment
Kajii et al. (47) 1966				+			5p-	
Laurent and Robert (52) 1966	III-1		+				5p-	All 3 cases are brothers
	III-5		+		+		5p-	Slight diffuse retinal atrophy
	III-6		+	+	+		5p-	Probable maternal translocation
Miller et al. (69) 1966b	Case 1	+	+	+	+	+	5p-	These cases also described by Breg et al. (8), Howard (41)
	Case 2	+	+	+	+	+	5p-	Family studies on patients 1 and 2 given by de Capoa et al. (10)
	Case 3	+	0	0	+	+	5p-	Further studies of Lejeune et al. (60) case 1; MacIntyre et al. (65); Wolf et al. (118)
	Case 5	+	0	0	0	+	5p-	
	Case 7	+	0	0	+	0	5p-	
	Case 8	+	+	+	+	0	5p-	
	Case 9	+	0	0	+	+	5p-	

TABLE 3-2 (continued)

Reference	Microcephaly	Epicanthus	Hypertelorism	Antimongoloid slant	Mongoloid slant	Strabismus	Optic atrophy	Chromosome abnormality	Familial translocation	Remarks
Milunsky and Chitham (71) 1966	+	+	+	+				5p–		Minimal strabismus
Neuhäuser and Lother (73) 1966	+	+	+	+		Con		5p–		Further description of Pfeiffer and Simon (84)
Olive et al. (80) 1966	+	+	+	+				5p–		
Passarge et al. (82) 1966								5p–		Chromosomal mosaicism; "typical cri-du-chat syndrome"
Reichelt et al. (89) 1966	+	+	+	+				5p–		
Schneegans et al. (98) 1966		(+)	+	+				5p–		

Silber et al. (103) 1966		+	+	+		+	5p-		
Steele et al. (107) 1966		+	+	+			5r		Chromosome mosaicism; microphthalmos
Turner et al. (112) 1966			+	+			5p-		Chromosomal mosaicism
Wolf et al. (119) 1966	+		+	+		Div	5p-	5/D2	
Zellweger (121) 1966		(+)	+		+		5p-		Chromosomal mosaicism
Buffoni et al. (9) 1967		+	+	+			5p-		
Koulischer et al. (50) 1967		3/3					5p-	5/D	1 case in this family previously described by Koulischer and Quersin (49)
Lichtenberger and Morineaud (61) 1967	+		+	+			5p-		Oval papillae
McGavin et al. (64) 1967		+		+		R. Con			Normal karyotype; bilateral megalocorneae; R. lower lacrimal punctum absent; acute L. dacryocystitis
Schlegel et al. (96) 1967	+		+		+		5p-		

TABLE 3-2 (continued)

Reference	Microcephaly	Epicanthus	Hypertelorism	Antimongoloid slant	Mongoloid slant	Strabismus	Optic atrophy	Chromosome abnormality	Familial translocation	Remarks
Schmid and Vischer (97) 1967	+		+	+		+		5p−		Normal fundi
Schroeder et al. (99) 1967		+	+					5p−		
Solitare (105) 1967	+	+	+					5p−		Primary micrencephaly; this case also described in de Capoa et al. (10)
Stahl et al. (106) 1967	2/2	2/2	2/2	2/2				5p−		
Vassela et al. (113) 1967	+	+		+				5p−		
Antich et al. (2) 1968		+	+					5p−		Chromosomal mosaicism
Bach et al. (4) 1968	+	0		0		0		5p−		

Gordon and Cooke (30) 1968	Case 1		+	+	+		Div	5p-		Previously described by McCracken and Gordon (63)
	Case 2		+	+				5p-		
	Case 3		0	0				5p-		
	Case 4		+					5p-		
	Case 5	+	+	+				5p-		
	Case 6	+		+				5p-		"Eyes not focusing"
Legros and van Michel (55) 1968		+	+	+				5p-		
Neuhäuser et al. (74) 1968			+	+	+		Con	5p-		Chromosomal mosaicism
Noel et al. (77) 1968	Case 1		+	+		+		5p-	+	(5p-;Gp+)mat; incomplete demyelination of the optic nerve
	Case 3		+	+				5p-	+	4 members in the sibship: 1 normal; 1 Gp+; 2 5p-
Olive et al. (79) 1968			+	+		+		5p-		
Petit et al. (83) 1968			+			+		5p-		Brushfield spots; complex mosaic; see text
Roca et al. (93) 1968		+	+		+			5p-		Chromosome mosaic

TABLE 3-2 (continued)

Reference	Microcephaly	Epicanthus	Hypertelorism	Antimongoloid slant	Mongoloid slant	Strabismus	Optic atrophy	Chromosome abnormality	Familial translocation	Remarks
Schwartz et al. (100) 1968		+	+			Div		5p-		Absent eyelashes; narrow palpebral fissures; microphthalmos
Sachsse et al. (95) 1969 Case 1			+					5p-		Chromosome mosaic
Case 2			+		+			5p-		Chromosome mosaic
Warburton et al. (114) 1969 Case 17	+	+	+	+		0		5p-		Cases 17 and 18 also described in de Capoa et al. (10)
Case 18	+	+	+	+		0		5p-		
Case 26	+	+	+	+		+		5p-		
Case 27	+	+	+	+		-		5p-		
Breg et al. (8) 1970 Case 1	+	+	+	+		Div	+	5p-		Cases 1, 2, 3, 5, 7, 9 previously described in Miller et al. (69). Case 15 also described in de Capoa et al. (10)
Case 2	+	+	+	+		Div	0	5p-		

Reference	Case								
Breg et al. (8) 1970	Case 3	+	0	0	+	Div	+	5p-	
	Case 5	+	0	0	R	Con	?	5p-	
	Case 7	+	0	0	0	0	+	5p-	
	Case 9	+	+	0	L	Div	+	5p-	
	Case 12	+	+	0	0	Div	+	5p-	Myopia
	Case 13	+	+	0	L	0	?	5p-	? Myopia
	Case 14	+	+	0	R	Div	?	5p-	? Myopia
	Case 15	+	+	0	+	Div	?	5p-	? Myopia
	Case 16	+	+	0	0	Div	0	5p-	Myopia
	Case 23	+	0	0	0	Div	?	5p-	? Myopia
	Case 25	+	+	+	0	Div	?	5p-	? Myopia
Jackson and Barr (44) 1970			+	+				5p-	Slight downward slant of the lateral canthus; 45, XY, 5-, 15-, t(5q15q); parents normal
Philip et al. (86) 1970				+				5p-	Maternal chromosomal mosaic
Altrogge et al. (1) 1971	Case 1			+	+	Div		5p-	Case 6 in this series is a 4p-; see also in Passarge et al. (see Table 3-1, ref. 81)
	Case 2		+	+	+			5p-	
	Case 3		+	+	+			5p-	

TABLE 3-2 (continued)

Reference	Microcephaly	Epicanthus	Hypertelorism	Antimongoloid slant	Mongoloid slant	Strabismus	Optic atrophy	Chromosome abnormality	Familial translocation	Remarks
Altrogge et al. (1) 1971 Case 4		+	+	+				5p-		
Case 5		+	+	+				5p-		
Catti and Schmid (12) 1971		+	+			Div				5(p-q+), Cr and a 15q+ chromosome
Ch'eng and Walz (13) 1971		+	+			Div		5p-		
Deminatti et al. (18) 1971	+	+	+					5r		
Grotsky et al. (31) 1971			+					5p-	4/5	Cataracts; maternal translocation
Niebuhr (76) 1971 Case 8	+	+				+		5p-		5p-, Yq+, Gp+, maternal Gp+
Case 10	+	+				+		5p-		

Niebuhr (76) 1971	Case 11 +				+				5p-	5p-, Dp+, paternal Dp+
	Case 12 +		+				+		5p-	
	Case 13 +		+						5p-	
	Case 18 +		+						5p-	
	Case 20 +		+						5p-	5p-, inv(Fp-q+), maternal (Fp-q+)
	Case 21 +		+				+		5p-	
Sedano et al. (101) 1971	Case 1 +		+		+		Con		5p-	
	Case 2 +		+		+		Div		5p-	
	Case 3 +		+		+				5p-	
Colover et al. (15) 1972				+					5p-	Very small deletion of 5p; nystagmus; neurological abnormalities
Howard (41) 1972	Case 1	+	0	+	+		Div	0	5p-	Cases 1-6 previously described by Breg et al. (8)
	Case 2	+	0	0	0		Div	+	5p-	Absent foveal reflex
	Case 3	+	0	+	0	0	Div	0	5p-	Tortuous arteries and veins, cases 1-6; deficient tears, cases 1-6
	Case 4	+	0	+	+	0	Div	+	5p-	Pupils supersensitive to 2.5% mecholyl
	Case 5	+	0	0	0	0	Div	0	5p-	
	Case 6	+	0	0	0	+	Div	0	5p-	

TABLE 3-2 (continued)

Reference	Microcephaly	Epicanthus	Hypertelorism	Antimongoloid slant	Mongoloid slant	Strabismus	Optic atrophy	Chromosome abnormality	Familial translocation	Remarks
Howard (41) 1972 Case 7	+	0		+	0	0	0	5p−		Hypotelorism; ptosis; presumed 5p−
Mann and Rafferty (66) 1972			+					5p−	5/11	Maternal translocation
Niebuhr (75) 1972		+	+	+		L Div				45, XX, 5-13−, dic+
Singh et al. (104) 1973		+	+							Maternal t(5p−;11q+)
Taillemite et al. (109) 1973	+		+							46, XX, 5p−, inv(4p;13q)
Kühner et al. (51) 1974	+	+	+	+		+		5p−		Nystagmus, normal fundi

REFERENCES

1. Altrogge, H. C., Hirth, L. and Goedde, H. W. Defizienz den kurzen Arme der Chromosomen der B-Gruppe (4p-:5p-). Z. Kinderheilkd. 110: 218-247 (1971).

2. Antich, J., Ribas-Mundo, M., Prats, J. and Roca, M. Cri-du-chat with chromosomal mosaicism. Lancet 1: 538 (1968).

3. Arias, D., Passarge, E., Engle, M. E. and German, J. Human chromosomal deletion: Two patients with the 4p- syndrome. J. Pediatr. 76: 82-88 (1970).

4. Bach, Ch., Gautier, M., Schaefer, P. and Moszer, M. La maladie du cri-du-chat. Une observation anatomoclinique. Ann. Pediatr. (Paris) 44: 339-343 (1968).

5. Berg, J. M., Delhanty, J. D. A., Faunch, J. A. and Ridler, M. A. C. Partial deletion of short arm of a chromosome of the 4-5 group (Denver) in an adult male. J. Ment. Defic. Res. 9: 219-228 (1965).

6. Bergman, S., Flodstrom, I. and Ansehn, S. "Cri du chat". Lancet 1: 768 (1965).

7. Bettecken, F., Reinwein, H., Künzer, W., Wolf, U. and Baitsch, H. Klinische und Genetische Untersuchungen en einem Patientem mit Cri-du-Chat-Syndrom. Dtsch. Med. Wochenschr. 90: 2008-2018 (1965).

8. Breg, W. R., Steele, M. W., Miller, O. J., Warburton, D., Capoa, A. de and Allderdice, P. W. The cri du chat syndrome in adolescents and adults: Clinical finding in 13 older patients with partial deletion of the short arm of chromosome No. 5 (5p-). J. Pediatr. 77: 782-791 (1970).

9. Buffoni, L., Monteverde, R. and Belotti, B. M. Studio clinico e genetico della malattia del "cri-du-chat". Minerva Pediatr. 19: 860-866 (1967).

10. Capoa, A. de, Warburton, D., Breg, W. R., Miller, D. A. and Miller, O. J. Translocation heterozygosis: A cause of 5 cases of the cri-du-chat syndrome and two cases with a duplication of chromosome number 5 in three families. Am. J. Hum. Genet. 19: 586-603 (1967).

11. Carter, R., Baker, E. and Hayman, D. Congenital malformations associated with a ring 4 chromosome. J. Med. Genet. 6: 224-227 (1969).

12. Catti, A. and Schmid, W. A pericentric inversion, 5p-q+ , and additional complex rearrangements in a case of cri-du-chat syndrome. Cytogenetics 10: 50-60 (1971).

13. Ch'eng, L. Y. and Walz, D. D. "Cri du chat" syndrome. Report of a case. Ohio State Med. J. 67: 237-240 (1971).

14. Citoler, P., Gropp, A. and Gulotta, F. Cytogenetische und pathologisch-anatomische Befunde bei 4p- Syndrom (Wolf-Syndrom). Beitr. Pathol. Anat. 143: 84-96 (1971).

15. Colover, J., Lucas, M., Comley, J. A. and Roe, A. M. Neurological abnormalities in the "Cri-du-chat" syndrome. J. Neurol. Neurosurg. Psychiatr. 35: 711-719 (1972).

16. Dallaire, L. A ring B chromosome in a female with multiple skeletal abnormalities. Birth Defects: Orig. Art. Ser. 5(pt 5): 114-116 (1969).

17. DeAlmeida, J. C. C., Gonzaga, M., Vieira, H., Brabosa, L. T., Abreu, M. C. and Barcinski, M. A. Partial deletion of the short arm of chromosome 5: "Le Cri du Chat" another example. Arq. Bras. Endocrinol. Metabol. 13: 183-192 (1964).

18. Deminatti, M., Cousin, J., Boutu, F., Savary, J. B., and Jacqueloot, N.: Maladie du cri du chat par chromosome 5 en anneau. In Proceedings of the Fourth International Congress of Human Genetics, J. de Grouchy, F. J. G. Ebling, I. Henderson, and J. François (Eds.), Amsterdam, Excerpta Medica (International Congress Series No. 233), 1971, p. 54.

19. Duke-Elder, W. S. System of Ophthalmology, Vol. 3 (pt. 2), London, Kimpton, 1964, p. 576.

20. Dumars, K. W., Jr., Gaskill, C. and Kitzmiller, N. Le cri du chat (crying cat). Am. J. Dis. Child. 108: 533-537 (1964).

21. Dyggve, H. V. and Mikkelsen, M. Partial deletion of the short arms of a chromosome of the 4-5 group (Denver). Arch. Dis. Child. 40: 82-85 (1965).

22. Engel, E., Hastings, C. P., Merrill, R. E., McFarland, B. S. and Nance, W. E. Apparent cri-du-chat and "antimongolism" in one patient. Lancet 1, 1130-1132 (1966)

23. Ford, E. H. R. Human Chromosomes. New York, Academic Press, Inc., 1973, p. 181 and p. 253.

24. François, J., Berger, R. and Saraux, H. Les Aberrations Chromosomiques en Ophtalmologie. Paris, Masson et Cie, 1972, pp. 147-165.

25. Fryns, J. P., Eggermont, E., Verresen, H. and Berghe, H. van den. The 4p- syndrome with a report of two new cases. Humangenetik 19: 99-109 (1973).

26. Genest, P., Tremblay, M. and Mortezai, M. Le syndrome du "cri-du-chat". Laval Med. 36: 319-327 (1965).

27. German, J. L., Lejeune, J., MacIntyre, M. N. and de Grouchy, J. Chromosomal autoradiography in the cri du chat syndrome. Cytogenetics 5: 347-352 (1964).

28. Giorgi, P. L., Ceccarelli, M. and Paci, A. Su un caso di sundrome del "cri-du-chat" con peculiari anomalie fenotipiche. Minerva pediatr. 17: 1972-1975 (1965).

29. Ginsberg, J. and Soukup, S. Ocular findings associated with ring B chromosomes. Am. J. Ophthalmol. 78: 624-629 (1974).

30. Gordon, R. R. and Cooke, P. Facial appearance in cri-du-chat syndrome. Dev. Med. Child Neurol. 10: 69-76 (1968).

31. Grotsky, H., Hsu, L. Y. F. and Hirschhorn, K. A case of cri-du-chat associated with cataracts and transmitted from a mother with a 4/5 translocation. J. Med. Genet. 8: 369-371 (1971).

32. de Grouchy, J. and Gabilan, J. C. Translocation 5/21-22 et syndrome du cri-du-chat. Ann. Genet. (Paris) 8: 31-38 (1965).

33. de Grouchy, J., Arthuis, M., Salmon, C., Lamy, M. and Thieffry, S. Le syndrome du cri-du-chat: Une nouvelle observation. Ann. Genet. (Paris) 7: 13-16 (1964).

34. Guthrie, R. D., Aase, J. M., Asper, A. C. and Smith, D. W. The 4p- syndrome. Am. J. Dis. Child. 122: 421-425 (1971).

35. Hamerton, J. L. Human Cytogenetics, Vol. II, New York, Academic Press, Inc., 1971, pp. 358-359.

36. Hecht, F. Ring 4 chromosome: Ring autosomes, lorelei of clinical-karyotype correlation and deletion mapping. Birth Defects: Orig. Art. Ser. 5(pt. 5): 106-113 (1969).

37. Hijmans, J. C. and Shearin, D. B. Partial deletion of short arms of chromosome No. 5. Am. J. Dis. Child. 109: 85-89 (1965).

38. Hirschhorn, K. and Cooper, H. L. Apparent deletion of short arms of one chromosome (4 or 5) in a child with defects of midline fusion. Human Chromosome Newsletter 4: 14 (1961).

39. Hirschhorn, K., Cooper, H. L. and Firschein, I. L. Deletion of short arms of chromosome 4-5 in a child with defects of midline fusion. Humangenetik 1: 479-482 (1965).

40. Holboth, N. and Mikkelsen, M. Maladie du cri-du-chat. Acta Obstet. Gynecol. Scand. 44: 572-578 (1965).

41. Howard, R. O. Ocular abnormalities in the cri-du-chat syndrome. Study of 7 patients. Am. J. Ophthalmol. 73: 949-954 (1972).

42. Hustinx, T. W. J. and Wijffels, J. C. H. M. "Cri-du-chat" syndrome. Lancet 2: 135 (1965a).

43. Hustinx, T. W. J. and Wijffels. J. C. H. M. "Cri-du-chat" syndrome. Maandschr. Kindergeneesskd. 33: 286-298 (1965b).

44. Jackson, L. and Barr, M. A 45,XY,5-,15-,t(5q15q) cri-du-chat child. J. Med. Genet. 7: 161-163 (1970).

45. Jalbert, P., Gilbert, Y., Léopold, Ph., Mouriquand, Cl. and Beaudoing, A. Syndrome d'Hallerman-Streiff-François. A propos d'une nouvelle observation associée à une anomalie caryotypique 4p-. Pediatrie 23: 703-705 (1968).

46. Judge, C. G., Garson, O. M., Pitt, D. B. and Sutherland, G. R. A girl with Wolf-Hirschhorn syndrome and mosaicism 46,XX/46,XX,4p-. J. Ment. Defic. Res. 18: 79-85 (1974).

47. Kajii, T., Homma, T., Oikawa, K., Furuyama, M. and Kawarazaki, T. Cri-du-chat syndrome. Arch. Dis. Child. 41: 97-101 (1966).

48. van Kempen, C. and Jongbloet, P. H. Partial deletion of the short arm of a chromosome No. 4. Maandschr. Kindergeneeskd. 35: 252-269 (1967).

49. Koulischer, L. and Quersin, Cl. La délétion des bras courts d'un chromosome du groupe 4-5. Acta Paediatr. Belg. 19: 167-179 (1965).

50. Koulischer, L., Petit, P. and Hayez-Delatte, F. La maladie du cri du chat - ségrégation familiale d'une translocation t(5p-;Dp+). Ann. Genet. (Paris) 10: 150-152 (1967).

51. Kühner, U., Büsse, M. and Buchinger, G. Cri-du-chat syndrome with an increased level of proline and threonine. Z. Kinderheilkd. 117: 259-264 (1974).

52. Laurent, C. and Robert, J. M. Etude génétique et clinique d'une famille de sept enfants dans laquelle trois sujets sont atteints de la maladie du "cri du chat". Ann. Genet. (Paris) 9: 112-122 (1966).

53. Leao, J. C., Bargman, G. J., Neu, R. L., Kajii, T. and Gardner, L. I. A new syndrome associated with partial deletion of short arms of chromosome No. 4. J. Am. Med. Assoc. 202: 434-436 (1967).

54. Leao, J. C., Neu, R. L. and Gardner, L. I. Hypospadias and other anomalies associated with partial deletion of short arms of chromosome No. 4. Lancet 1: 493 (1966).

55. Legros, J. and Michel, Cl. van. Analyse de la voix dans un cas de "maladie du cri-du-chat". Ann. Genet. (Paris) 11: 59-61 (1968).

56. Lejeune, J., Gautier, M., Lafourcade, J., Berger, R. and Turpin, R. Délétion partielle du bras court du chromosome 5: Cinquième cas du syndrome du cri du chat. Ann. Genet. (Paris) 7: 7-10 (1964a).

57. Lejeune, J., Lafourcade, J., Berger, R. and Réthoré, M.-O. Maladie du cri-du-chat et sa réciproque. Ann. Genet. (Paris) 8: 11-15 (1965).

58. Lejeune, J., Lafourcade, J., Berger, R. and Turpin, R. Ségrégation familiale d'une translocation 5-13 déterminant une monosomie et une trisomie partielle du bras court du chromosome 5: Maladie du "cri-du-chat" et sa "réciproque". C. R. Acad. Sci. [D] (Paris), 258: 5767-5770 (1964b).

59. Lejeune, J., Lafourcade, J., Berger, R., Vialatte, J., Boeswillwald, M., Seringe, P. and Turpin, R. Trois cas de délétion partielle du bras court d'un chromosome 5. C. R. Acad. Sci. [D] (Paris) 257: 3098-3102 (1963).

60. Lejeune, J., Lafourcade, J., Gautier, M., de Grouchy, J., Berger, R., Solmon, C. and Turpin, R. Délétion partielle du bras court du chromosome 5: Individualisation d'un nouvel état morbide. Sem. Hop. Paris 40: 1069-1079 (1964c).

61. Lichtenberger, M. and Morineaud, J. -P. Un cas de délétion du bras court du chromosome 5 au Vietnam (syndrome du cri-du-chat). Ann. Pediatr. 44: 742-755 (1967).

62. Lindenbaum, R. H. and Butler, L. J. Child with multiple anomalies and a group B (4-5) long arm deletion. Arch. Dis. Child. 46: 99-101 (1971).

63. McCracken, J. S. and Gordon, R. R. "Cri du chat" syndrome: A new clinical and cytogenetic entity. Lancet 1: 23-25 (1965).

64. McGavin, D. D. M., Cant, J. S., Ferguson-Smith, M. A. and Ellis, P. M. The cri-du-chat syndrome with an apparently normal karyotype. Lancet 2: 326-330 (1967).

65. MacIntyre, M. N., Staples, W. I., LaPolla, J. and Hempel, J. M. The "cat-cry syndrome". Am. J. Dis. Child. 108: 538-542 (1964).

66. Mann, J. and Rafferty, J. H. Cri-du-chat syndrome combined with partial C group trisomy. J. Med. Genet. 9: 289-292 (1972).

67. Miller, O. J., Breg, W. R., Warburton, D., Miller, D. A., de Capoa, A. and Chutorian, A. M. Deleted late-replicating chromosome 4/5. Lancet 2: 105 (1966a).

68. Miller, O. J., Breg, W. R., Warburton, D., Miller, D. A., de Capoa, A., Allderdice, P. W., Davis, J., Klinger, H. P., McGilvray, E. and Allen, F. H. Jr. Partial deletion of the short arm of chromosome No. 4 (4p-): Clinical studies in 5 unrelated patients. J. Pediatr. 77: 792-801 (1970).

69. Miller, O. J., Breg, W. R., Warburton, D., Miller, D. A., Firschein, I. L. and Hirschhorn, K. Alternative DNA replication patterns associated with long arm length of chromosomes 4 and 5 in the cri-du-chat syndrome. Cytogenetics 5: 137-151 (1966b).

70. Miller, O. J., Warburton, D. and Breg, W. R. Deletions of Group B chromosomes. Birth Defects: Orig. Art. Ser. 5(pt. 5): 100-105 (1969).

71. Milunsky, A. and Chitham, R. G. The cri du chat syndrome. J. Ment. Defic. Res. 10: 153-157 (1966).

72. Moyer, F. G. The Wolf-Hirschhorn (4p-) syndrome. Birth Defects: Orig. Art. Ser. 7(pt. 7): 314-317 (1971).

73. Neuhäuser, G. and Lother, K. Das Katzenschrei-Syndrom. Monatsschr. Kinderheilkd. 14: 278-281 (1966).

74. Neuhäuser, G., Singer, H. and Zang, K. D. Cri du chat Syndrom mit Chromosomen-mosaik 46, XY/46, XY, 5p-. Humangenetik 5: 315-320 (1968).

75. Niebuhr, E. A 45, XX, 5-, 13-, dic+ karyotype in a case of cri-du-chat syndrome. Cytogenetics 11: 165-177 (1972).

76. Niebuhr, E. The cat cry syndrome (5p-) in adolescents and adults. J. Ment. Defic. Res. 15: 277-291 (1971).

77. Noël, B., Quack, B. and Thiriet, M. Ségrégation d'une translocation balancée t(5p-;Gp+). Ann. Genet. (Paris) 11: 247-252 (1968).

78. Ockey, C. H., Feldman, G. V., Macaulay, M. E. and Delaney, M. J. A large deletion of the long arm of chromosome no. 4 in a child with limb abnormalities. Arch. Dis. Child. 42: 428-434 (1967).

79. Olive, D., Gilgenkrantz, S., Cabrol, C. and Wolkowicz, M. W. Maladie du cri du chat (5p-) avec gueule de loup. Pediatrie 23: 795-800 (1968).

80. Olive, D., Gilgenkrantz, S., Kiffer, B., Hendryckx, J. and Pierson, M. La maladie du cri du chat. Arch. Fr. Pediatr. 23: 976 (1966).

81. Passarge, E., Altrogge, H. C. and Rudiger, R. A. Human chromosomal deficiency: The 4p- syndrome. Humangenetik 10: 51-57 (1970).

82. Passarge, E., Jarrett, T. E., Smith, L. B. and Soukup, S. W. Mosaicism for a deleted chromosome 5? Human Chromosome Newsletter 19: 30 (1966).

83. Petit, P., Maurus, R., Richard, J. and Koulischer, L. Chromosome du cri du chat chez un trisomique 21 leucémique. Ann. Genet. (Paris) 11: 125-128 (1968).

84. Pfeiffer, R. A. and Simon, H. A. "Cri du chat" ein neues Missbildungssyndrom als Folge einer Chromosomenaberration. Munch. Med. Wochenschr. 107: 2669-2674 (1965).

85. Pfeiffer, R. A. Neue Dokumentation zur Abgrenzung eines Syndroms der Deletion des kurzen Arms eines Chromosoms Nr 4. Z. Kinderheilkd. 102: 49-61 (1968).

86. Philip, J., Brandt, N. J., Friis-Hansen, B., Mikkelsen, M. and Tygstrup, I. A deleted B chromosome in a mosaic mother and her cri du chat progeny. J. Med. Genet. 7: 33-36 (1970).

87. Polani, P. E. Autosomal imbalance and its syndromes excluding Down's. Br. Med. Bull. 25: 81-93 (1969).

88. Punnett, H. H., Carpenter, G. G. and Di George, A. M. Deletion of short arm of chromosome 5. Lancet 2: 588 (1964).

89. Reichelt, W., Voigt, G. and Zernahle, K. Zur Symptomatik und Cytogenetik eines weiteren Falles mit Katzenschreisyndrom. Monatsschr. Kinderheilkd. 114: 592-596 (1966).

90. Reinwein, H. and Wolf, U. "Cri du chat" with 5/D translocation. Lancet 2: 797 (1965).

91. Ricci, N., Ventimiglia, B., Dallapiccola, B., Franceschini, F. and Preto, G. "Cri du chat" syndrome. Lancet 1: 1278 (1965).

92. Ricci, N., Ventimiglia, B., Dallapiccola, B., Franceschini, F. and Preto, G. La sindrome da delezione del bracio corto di un cromosoma 4-5. Acta Genet. Med. Gemellol. (Roma), 15: 36-50 (1966).

93. Roca, M., Antich, J., Ribas-Mundo, M. and Prats, J. Sindrome del "cri du chat" con mosaico cromosomico. Med. Clin. 50: 180-184 (1968).

94. Rohde, R. A. and Tompkins, R. "Cri du chat" due to a ring-B (5) chromosome. Lancet 2: 1075 (1965).

95. Sachsse, W., Schmidt, E. and Hellmich, E. Beobachtungen zum Mosaizimus beim Cri du Chat Syndrom. Humangenetik 8: 71-74 (1969).

96. Schlegel, R. J., Neu, R. L., Leao, J. C., Reiss, J. A., Nolan, T. B. and Gardner, L. I. Cri du chat syndrome in a 10 year old girl with deletion of the short arms of chromosome number 5. Helv. Paediatr. Acta 22: 2-12 (1967).

97. Schmid, W. and Vischer, D. Cri-du-chat syndrome. Helv. Paediatr. Acta 22: 22-27 (1967).

98. Schneegans, E., Rohmer, A., Levy-Silagy, J., Rumpler, Y., Ruch, J. -V. and Gerlinger, P. Un cas de maladie du cri du chat, délétion partielle du bras court du chromosome 5. Pediatrie 21: 823-834 (1966).

99. Schroeder, H. -J., Schleiermacher, E., Schroeder, T. M., Bauer, H., Richter, C. and Schwenk, J. Zur klinischen Differentialdiagnose des Cri du Chat Syndrom. Humangenetik 4: 294-304 (1967).

100. Schwartz, J. J., Chinitz, J. and Kushnick, T. Cri du chat syndrome with additional physical findings of trisomy 17-18. Lancet 88: 303-305 (1968).

101. Sedano, H. O., Look, R. A., Carter, C. and Cohen, M. M. Jr. B-group short arm deletion syndromes. Birth Defects: Orig. Art. Ser. 7(pt. 7): 89-97 (1971).

102. Sidbury, J. B. Jr., Schmickel, R. D. and Gray, M. Findings in a patient with apparent deletion of short arms on one of the B group chromosomes. J. Pediatr. 65: 1098 (1964).

103. Silber, D. L., Engel, E. and Merrill, R. E. So-called "Cri du chat" syndrome. Am. J. Ment. Defic. Res. 71: 486-491 (1966).

104. Singh, D. N., Osborne, R. A. and Wiscovitch, R. A. Transmission of the cri-du-chat syndrome from a maternal balanced translocation carrier t(5p-:11q+). Humangenetik 20: 361-365 (1973).

105. Solitare, G. B. The cri du chat syndrome: Neuropathologic observations. J. Ment. Defic. Res. 11: 267-277 (1967).

106. Stahl, A., Louchet, E., Giraud, F., Soulayrol, R., Hartung, M., Brusquet, Y. and Bernard, P. -J. Le syndrome de délétion des bras courts d'un chromosome 5 (maladie du cri du chat). A propos de deux observations. Arch. Fr. Pediatr. 24: 716 (1967).

107. Steele, M. W., Breg, W. R., Eidelman, A. I., Lion, D. T. and Terzakis, T. A. A B-group ring chromosome with mosaicism in a newborn with cri-du-chat syndrome. Cytogenetics 5: 419-429 (1966).

108. Šubrt, I., Blēhová, B. and Sedláčková, E. Mewing cry in a child with the partial deletion of the short arm of chromosome No. 4. Humangenetik 8: 242-248 (1969).

109. Taillemite, J. -L., Baheux-Morlier, G., Cathelineau, L. and Roux, C. Maladie du cri du chat associée à un remaniement chromosomique complexe chez un jumeau dizygote. Ann. Genet. (Paris) 16: 127-130 (1973).

110. Taylor, A. I. Patau's, Edwards' and Cri du chat syndromes: A tabulated summary of current findings. Dev. Med. Child Neurol. 9: 78-86 (1967).

111. Taylor, A. I., Challacombe, D. N. and Howlett, R. M. Short-arm deletion, chromosome 4 (4p-), a syndrome? Ann. Hum. Genet. 34, 137-144 (1970).

112. Turner, J. H., Bass, L. W. and Kaplan, S. Chromosome mosaicism in a child with features characteristic of the 'cat cry' syndrome. J. Med. Genet. 3: 66-69 (1966).

113. Vassella, F., Joss, E., Luchsinger, R., Dubois, C., Gloor, R. and Wiesmann, U. Cri-du-Chat-Syndrom. Helv. Paediatr. Acta 22: 13-21 (1967).

114. Warburton, D., Miller, D. A., Miller, O. J., Allderdice, P. W. and Capoa, A. de. Detection of minute deletions in human karyotypes. Cytogenetics 8: 97-108 (1969).

115. Wilcock, A. R., Adams, F. G., Cooke, P. and Gordon, R. R. Deletion of short arm of no 4 (4p-) - A detailed case report. J. Med. Genet. 7: 171-176 (1970).

116. Wilson, M. G., Towner, J. W. and Negus, L. D. Wolf-Hirschhorn syndrome associated with an unusual abnormality of chromosome No 4. J. Med. Genet. 7: 164-170 (1970).

117. Wolf, U. and Reinwein, H. Klinische und cytogenetische Differentialdiagnose der Defizienzen an den kurzen Armen der B-Chromosomen. Z. Kinderheilkd. 98: 235-245 (1967).

118. Wolf, U., Porsch, R., Baitsche, H. and Reinwein, H. Deletion on short arms of a B-chromosome without "cri-du-chat" syndrome. Lancet 1: 769 (1965a).

119. Wolf, U., Reinwein, H., Gey, W. and Klose, J. Cri-du-Chat Syndrom mit Translokation 5/D_2. Humangenetik 2: 63-67 (1966).

120. Wolf, U., Reinwein, H., Porsch, R., Schröter, R. and Baitsch, H. Defizienz an den Kurzen Armen eines Chromosoms Nr 4. Humangenetik 1: 397-413 (1965b).

121. Zellweger, H. Cri du chat with chromosomal mosaicism. Lancet 2: 57 (1966).

CHAPTER 4

DEFICIENCIES OF GROUP C CHROMOSOMES

I. OCULAR ABNORMALITIES

This is a small group of cases having various abnormalities such as microcephaly, hypertelorism, and epicanthus which are not specific to group C chromosomes. However, a few cases should be discussed. Gacs et al., (7) described a family where the proband had a twin sister who had microphthalmos, cleft lip and palate with facial asymmetry. There was a high incidence of congenital malformations and spontaneous abortions on the maternal side of the family, which might have been explained by a familial translocation, and family members who were tested, including the mother, all had normal karyotypes.

There are several cases where the parents all had normal karyotypes (4, 5, 8, 11), while in other cases, the parents were not tested (3, 14). However, the mother of case 1 described by Alfi et al., (1) did have a balanced translocation t(9p-;16q+), and consequently her daughter inherited the deficient 9p- chromosome. The authors also mention that the phenotype in their two cases is the countertype of 9p+. There are 17 cases of 9p+ in this review, but it is not yet possible to compare them with these two cases of 9p- and to come to a valid conclusion on ophthalmological grounds alone.

The case of Ladda et al., (10) is also interesting and is mentioned in the section on balanced translocations, where sophisticated methods of chromosome analysis were used to detect a deletion of chromosome 8 in a subject with the syndrome of aniridia and Wilm's tumour. The association of a chromosomal abnormality in this syndrome is not of special significance, but the technique used is of great interest since the chromosomal abnormality appeared at first to be a balanced translocation t(8p+;11q-).

TABLE 4-1

GROUP C CHROMOSOMES: DELETIONS AND RINGS

Author	Ocular abnormalities	Chromosome abnormality
Atkins et al. (2) 1966	Case 1: low forehead with prominent ridge over the frontal suture; small epicanthic folds	Cr
Butler et al. (3) 1967	Thick eyebrows and joined above bridge of the nose; eyes prominent, otherwise normal	Cr
Wurster et al. (15) 1969	Eyes spaced wide apart; epicanthus	Cr
Gacs et al. (7) 1970	Proband had microcephaly, mongoloid slant, hypertelorism, epicanthus, horizontal nystagmus, discoloured papillae	Cr
Kistenmacher and Punnet (9) 1970	Case 1: bushy eyebrows, hypertelorism, internal strabismus on the left, mongoloid slant, epicanthus	? 9r
Michiels et al. (12) 1972	Case 1: narrow palpebral fissures, antimongoloid slant, convergent strabismus, normal papillae but diffuse pigmentation of the retina, particularly in the macular region	Cr
Moore et al. (13) 1973	Microcephaly, microphthalmos, epicanthus, depressed nasal bridge	6r
Fraisse et al. (5) 1974	Microcephaly, slight epicanthus, no hypertelorism, slight exophthalmos	9r
de Grouchy et al. (8) 1968	Deeply depressed nasal bridge, protruding eyes, hypertrophied eyelids, epicanthus	? 6p-
Laurent et al. (11) 1968	Microcephaly, antimongoloid slant	10p- or 12p-

TABLE 4-1 (continued)

Author	Ocular abnormalities	Chromosome abnormality
Elliott et al. (4) 1970	Narrow skull, prominent occiput, shallow orbits	?10p-
Moore and Engel (14) 1970	Case 4: hypertelorism, narrow palpebral fissures, exotropia and nystagmus at right lateral gaze	Cp-
Alfi et al. (1) 1973	Case 1: prominent forehead, flattened occiput, slight mongoloid slant, mild hypertelorism; case 2: wide-set eyes	9p-
Friedrich et al. (6) 1974	Retracted supraorbital margins, mongoloid slant, ptosis, Brushfield spots	7q-
Ladda et al. (10) 1974	Bilateral aniridia, glaucoma, megacornea, anterior polar cataracts, unilateral Wilm's tumour	8p-

REFERENCES

1. Alfi, O., Donnell, G. N., Crandall, B. F., Derencsenyi, A. and Menon, R. Deletion of the short arm of chromosome No 9 (46, 9p-): A new deletion syndrome. Ann. Genet. (Paris) 16: 17-22 (1973).

2. Atkins, L., Pant, S. S., Hazard, G. W. and Ovellette, E. M. Two cases of C group ring autosome. Ann. Hum. Genet. 30: 1-5 (1966).

3. Butler, L. J., France, N. E. and Jacoby, N. M. An infant with multiple congenital anomalies and a ring chromosome in group C (X-6-12). J. Med. Genet. 4: 295-298 (1967).

4. Elliott, D., Thomas, G. H., Condron, C. J., Khuri, N. and Richardson, F. C-group chromosome abnormality (?10p-). Am. J. Dis. Child. 119: 72-73 (1970).

5. Fraisse, J., Lauras, B., Ooghe, M. J., Freycon, F. and Réthoré, M. -O. A propos d'un cas de chromosome 9 en anneau. Identification par dénaturation ménagée. Ann. Genet. (Paris) 17: 175-180 (1974).

6. Friedrich, U., Lyngbye, T. and Øster, J. A girl with karyotype 46, XX, del(7)(qter-p15). Humangenetik 26: 161-165 (1974).

7. Gacs, G., Schuler, D. and Sellyei, M. Familial occurrence of congenital malformations and ring chromosome (46, XX, Cr). J. Med. Genet. 7: 177-179 (1970).

8. Grouchy, J. de, Veslot, J., Bonnette, J. and Roidot, M. A case of ?6p- chromosomal aberration. Am. J. Dis. Child. 115: 93-99 (1968).

9. Kistenmacher, M. L. and Punnett, H. H. Comparative behaviour of ring chromosomes. Am. J. Hum. Genet. 22: 304-318 (1970).

10. Ladda, R., Atkins, L., Littlefield, J., Neurath, P. and Marimuthu, K. M. Computer-assisted analysis of chromosomal abnormalities: Detection of a deletion in aniridia/Wilm's tumor syndrome. Science 185: 784-787 (1974).

11. Laurent, C., Nivelon, A., Hartman and Guerrier, G. Monosomie partielle d'un chromosome du groupe C: (Cp-). Ann. Genet. (Paris) 11: 231-235 (1968).

12. Michiels, J., Stanescu, B. and Rommel, J. Trouble pigmentaire de la rêtine en cas d'aberration chromosomique. Bull. Mem. Soc. Fr. Ophtalmol. 85: 39-44 (1972).

13. Moore, C. M., Heller, R. H. and Thomas, G. H. Developmental abnormalities associated with a ring chromosome 6. J. Med. Genet. 10: 299-302 (1973).

14. Moore, M. K. and Engel, E. Clinical, cytogenetic, and autoradiographic studies in 10 cases with rare chromosome disorders. II. Cases 3, 4, and 5. Ann. Genet. (Paris) 13: 129-134 (1970).

15. Wurster, D., Pomeroy, J., Benirschke, K. and Hoefnagel, D. Mental deficiency and malformations in a boy with a group C ring chromosome: 46, XY, Cr. J. Ment. Defic. Res. 13: 184-190 (1969).

CHAPTER 5

DEFICIENCIES OF GROUP D CHROMOSOMES

I. GROUP D RINGS

The behaviour of ring chromosomes in human beings was first described by Wang et al. (56), one of his cases having a ring D and the other a ring E chromosome. There are 34 cases of ocular abnormalities associated with a Dr anomaly as compared with 18 such cases with a Dq- anomaly, but this by no means indicates the relative incidence of the two syndromes. The two karyotypes do produce syndromes which are similar; these are compared by Neibuhr and Ottosen (35), who divided 50 cases of Dr and Dq- from the literature into three groups, according to their clinical features.

A. Phenotype

In a recent review of the literature on Dr (5), the incidence of the most common abnormalities is given as psychomotor retardation (100%), low birth weight (90%), microcephaly (78%), anomalous ears (67%), micrognathia (43%), congenital heart disease (48%), and central nervous system anomalies such as arhinencephaly and trigonencephaly (39%) (Fig. 5-1). The incidence of absent or hypoplastic thumbs is not given in this review, but in the review of Neibuhr and Ottosen (35) they occur in 16 out of 50 cases of Dq- and Dr combined, but only in six out of 33 cases of Dr.

B. Ocular Abnormalities

In all, 34 cases of Dr abnormality associated with ocular findings have been found in the literature and these are given in Table 5-1. It should

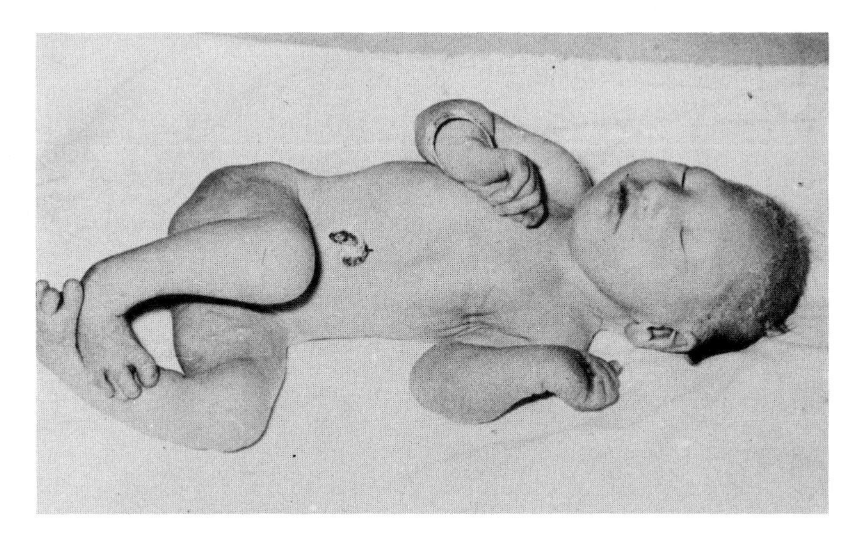

FIG. 5-1. A case of 13r with microcephaly, hypoplastic thumbs, and abnormal toes. (By permission of Dr. Renata Lax and Dr. Michael Ridler)

be noted that Bloom et al., (7) and Gerald et al., (13) described the same patient, the clinical description is given in one paper (13) while the anomalous inheritance of haptoglobin is discussed in the other paper (7). The most common findings are hypertelorism (19 cases), epicanthus (21 cases), and microphthalmos (6 cases). In addition, ptosis was seen in two cases (1, 27), antimongoloid slant in two (2, 34), mongoloid slant in five (9, 13, 24, 33, 44) and narrow or short palpebral fissures in four cases (3, 9, 15, 44). Strabismus, where it occurred, was convergent in three cases (24, 28, 51), divergent in one (55), and alternating in one (9). Nystagmus was only reported twice (5, 49). Iris coloboma was found in four cases but not as an isolated finding, being associated with (a) a 1 mm opacity in the anterior segment of the left eye with bilateral retinal colobomas (49); (b) with a lens coloboma (28); (c) with hypoplastic optic discs and retinal pigment mottling (24); and (d) with a choroidal coloboma (44, 48). Hypoplastic optic discs have also been described (22) as well as a cherry-red spot at the macula (51). Réthoré et al., (44) and Saraux et al., (48) have described the same patient, and there is a further description of this case by François et al., (11), all three reports with detailed ocular pathology. The essential features of this case are the absence of closure of the fetal cleft with the formation of a choroidal coloboma, and with a small colobomatous retinal cyst and nodule of cartilage behind the cyst. Also present were a cataract, persistence of tunica vasculosa lentis and an immature fetal-type angle of the anterior chamber.

The only other description of ocular pathology in Dr is that of Bilchik et al. (5), who described bilateral colobomas of the posterior poles and cataracts. New observations for this syndrome are the unusual structure of the iris, the recessed appearance of the angle of the anterior chamber, and the endothelial cell proliferation on the inner retina.

The case described by Grace et al. (16), in which retinoblastoma was found, may be considered as atypical for a Dr abnormality since it belongs more to the Dq- group where 10 cases of retinoblastoma have been described. On the other hand, it may be considered as a good example of the "overlapping" of clinical signs which one would expect theoretically in Dq- and Dr. It has been suggested that since this patient did not show the typical signs of the Dr syndrome, the presence of retinoblastoma may be accidental or may be due to the effect of an unstable ring (35).

TABLE 5-1

Group D Chromosomes: Dr

Reference	Microcephaly	Ptosis	Antimongoloid Slant	Mongoloid Slant	Narrow Palpebral Fissures	Hypertelorism	Epicanthus	Microphthalmos	Strabismus	Iris Coloboma	Chromosome Abnormality	Remarks
Wang et al. (56) 1962 Case 1							+				Dr	
Bain et al. (4) 1963								+			Dr	
Adams et al. (1) 1965		+				+	+				Dr	
Gerald et al. (13) 1967				+							Dr	Haptoglobin inheritance discussed in Bloom et al. (7)
Neimann et al. (34) 1967			+			+					Dr	
Sparkes et al. (49) 1967							+			+	14r	Colobomas of retinae. 1 mm opacity in anterior chamber LE, nystagmus

Teplitz et al. (51) 1967								L Con	Dr	RE: cherry-red spot on the macula
Lejeune et al. (28) 1968 Case 1							+			RE: white retrocorneal mass
Case 2	+							R Con +		Lens coloboma, no cataract
Case 3						+				Iris heterochromia, all 3 cases are 46, XY, -D, +Dr/45, XY, -D
Masterson et al. (30) 1968	+				+				Dr	
Allderdice et al. (2) 1969 Case 2			+		+				13r	
Ayraud and Szepetowski (3) 1969	+			+		+			Dr	
Juberg et al. (22) 1969		(+)			+	(+)			Dr	Optic discs small and dark
Mikkelsen and Niebuhr (32) 1969	+				+	+			13r	Familial t(13q14q)
Tolksdorf et al. (54) 1969	+				+	+			Dr	Thin eyebrows
Varela and Sternberg (55) 1969 Case 2					+			L Div	13r	

TABLE 5-1 (continued)

Reference	Microcephaly	Ptosis	Antimongoloid Slant	Mongoloid Slant	Narrow Palpebral Fissures	Hypertelorism	Epicanthus	Microphthalmos	Strabismus	Iris Coloboma	Chromosome Abnormality	Remarks
Biles et al. (6) 1970						+		(+)			13r	
Coffin and Wilson (9) 1970				+	+	+	+		Alt		13r	
Kistenmacher and Punnett (24) 1970 Case 2				+		+	+		R Con	+	Dr	Hypoplastic optic discs with temporal pallor, venous tortuosity, retinal pigment mottling
Morič̌-Petrovič et al. (33) 1970				+		+					13r	
Réthoré et al. (44) 1970				+	+	+	+	+		+	13r	Same case as Saraux et al. (48). Ocular pathology: choroido-scleral coloboma, small colobomatous retinal cyst, nodule of

Saraux et al. (48) 1970				+	+	+	+	13r	cartilage behind the cyst, cataract, tunica vasculosa lentis, immature type angle, no retinal dysplasia
Tolksdorf et al. (53) 1970 Case 1	+			+	+			Dr	
Case 2	+			+	+			Dr	
Gilgenkrantz et al. (15) 1971			+	+	+			14r	46, XX, Dr/45, XX, D-
Grace et al. (16) 1971 Case 2								Dr	Retinoblastoma
Hollowell et al. (18) 1971	+				+			13r	Almond-shaped eyes
Lehrke et al. (27) 1971								Dr	Possible iris asymmetry
Bilchik et al. (5) 1972					+	+		Dr	Blepharophimosis; posterior pole colobomata; anterior uveal tract hypoplasia; cataract; ocular pathology; nystagmus
Picciano et al. (42) 1972					+			Dr	
Salamanca et al. (46) 1972	+	+		+	+			13r	

TABLE 5-1 (continued)

Reference	Microcephaly	Ptosis	Antimongoloid Slant	Mongoloid Slant	Narrow Palpebral Fissures	Hypertelorism	Epicanthus	Microphthalmos	Strabismus	Iris Coloboma	Chromosome Abnormality	Remarks
Niebuhr and Ottosen (35) 1973						+	+				13r	Proptosis due to hypoplastic supraorbital and infraorbital margins
Fryns et al. (12) 1974	+					+	+	+			13r	Severely retarded 14-yr-old
Hoo et al. (19) 1974	+						+				13r	No eye abnormalities

II. GROUP D DELETIONS

Various attempts have been made to define a 13q-, 14q- and a 15q- syndrome (11, 26); there has, however, been only one case of 15q- (26) and one case of a possible 14q- or 15q- (8) where in both instances autoradiography was done, so that it is difficult on this basis to give a valid description of three separate syndromes. Now that accurate identification is possible with banding, it is apparent that the deletion involves chromosome 13 and only occasionally, 14 and 15. The original case of Lele, Penrose, and Stallard (29) is now known to be 13q- rather than 15q-.

A. Phenotype

The clinical features which have been described in the 13q- syndrome are mental retardation (microcephaly), broad prominent nasal bridge, abnormalities of the ears, facial asymmetry, micrognathia, protruding maxilla, short neck with folds, congenital heart disease, imperforate anus, hypospadias, bifid scrotum, pelvic girdle anomalies, foot and toe anomalies, absent or hypoplastic thumbs, shortened 4th and/or 5th finger, as well as ocular abnormalities.

B. Ocular Abnormalities

There are 16 cases of known 13q-, two with 15q- and two with Dp-, that are associated with ocular abnormalities which have been listed in Table 5-2. The most common findings are ptosis, antimongoloid slant, hypertelorism, epicanthus, and microphthalmos. Strabismus, where present, is usually convergent, colobomas of the iris are usually bilateral, and in two cases, the choroid was involved as well.

The ocular abnormality in this syndrome which is of greatest interest to ophthalmologists is retinoblastoma of which ten cases associated with 13q- and one with 13r have been published to date. It is now accepted that the deleted chromosome is chromosome 13, although some earlier reports suggested otherwise. Most of these cases were published before autoradiography or banding, and it is probable that if subsequent studies could be done on patients previously said to be a 14q- or a 15q-, it would be found that they were in fact 13q-. This was actually done by Wilson et al. (59) who identified the deleted chromosome as 13 by banding studies in a patient whom they had previously described in 1969 as 14q- (58). Case 2 of Grace et al. (16) is included in this group, although it is associated with a ring D chromosome. One would expect a ring chromosome to produce symptoms in common with those produced by a short arm and a long arm deletion.

The association of retinoblastoma and mental retardation with a 13q- deletion is recognized as a clinical entity. There have been, however, two studies of chromosomes in patients with retinoblastoma, that of Pruett and Atkins (43) and that of Wiener, Reese, and Hyman (57) where in a total of 38 patients from both studies, no chromosomal abnormality was detected. Jensen and Miller (21), however, made a large epidemiological study of a total of 1,346 cases, and of those, six had anomalies of other organ systems suggesting a D-deletion syndrome, but in this large series chromosome studies were not performed. There is a case of retinoblastoma with a 48,XXY, 21+ karyotype (45), and another with a 48,XXX, 21+ karyotype (10).

Banding studies have been carried out by various groups of investigators (20, 25, 40, 59). From these studies and from a study of the excellent review by Niebuhr and Ottosen (35), it is possible to suggest a model of a deletion map of chromosome 13 (Fig. 5-2). Loss of the terminal portion, bands q33 and q34, leads to microcephaly, abnormal ears, and mental retardation. Loss of bands q31 and q32 leads to abnormally small eyes, genital malformations, hypoplastic or absent thumbs, and other abnormalities. If the dense band q21 is missing, there is usually retinoblastoma and mental retardation. Wilson et al. (59) were of the opinion that in their patient with retinoblastoma, the missing band was q31 rather than q21, but in their patient with retinoblastoma, Howard et al. (20) found the missing band to be q21. There can be some difficulty in distinguishing q21 and q31 as they are somewhat similar in size. The above hypothetical model may have to be modified in the light of future findings.

It is to be hoped that in any new cases of retinoblastoma with a Dq- deletion that are published, there will be banding studies to identify the D chromosome, and it would be valuable to study the chromosomes of the parents in order to exclude a translocation. It would also be interesting to study the chromosomes of retinoblastoma tumour cells. At the time of writing, there have been numerous studies of tumour cells in retinoblastoma, but not of their chromosomes. The chromosomes in cultured tumour cells are often abnormal since new mutations occur which are not suppressed by immunological or other mechanisms in the body. It may be possible to find clones, which are groups of genetically identical cells originating from a single cell that has undergone a chromosome mutation. If 13q- clones were found in tumour cells

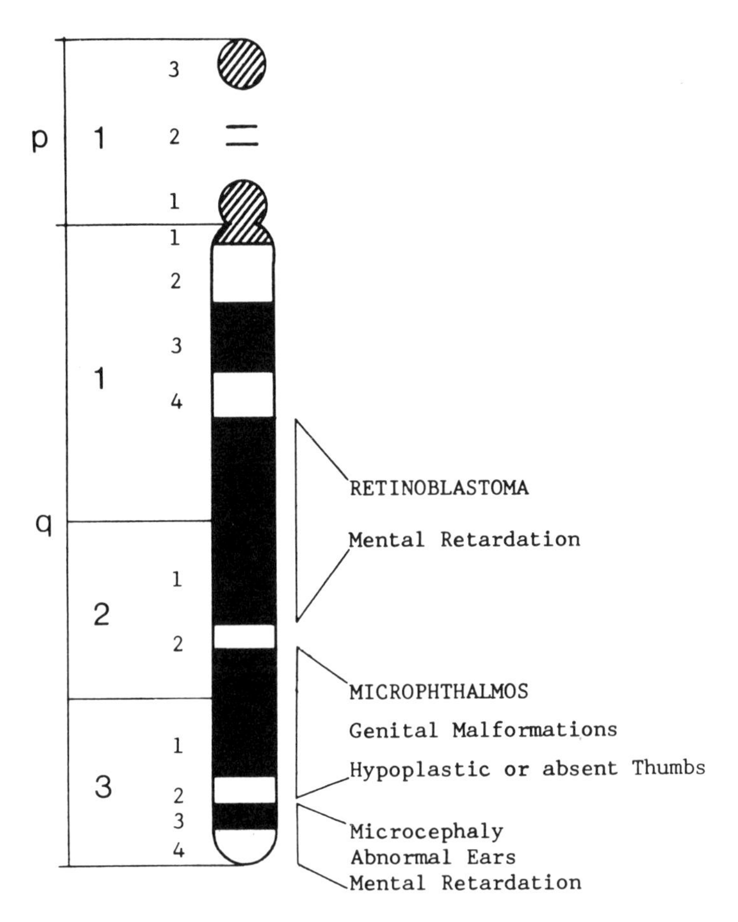

FIG. 5-2. Diagram of chromosome 13 with an indication of the different syndromes possibly resulting from deletions of different areas of the long arm.

of retinoblastoma, it would suggest that this form of retinoblastoma is due to the mutation of a gene localised on 13q.

Another aspect of this topic is the fact that in the ten cases of retinoblastoma, seven are bilateral, and bilateral retinoblastoma is thought to have a single gene dominant mode of inheritance. It has been suggested that there may be a gene for normal retinal formation located on chromosome 13, and, if this is so, it is reasonable to assume that it is on the segment of the long arm which is deleted in cases of retinoblastoma.

The ocular findings in patients with partial monosomy D are compared in Table 5-3. Certain abnormalities are common to both Dr and Dq- such as hypertelorism, epicanthus, mongoloid or antimongoloid slant, but these are also commonly seen in every chromosomal abnormality and are probably polygenic in origin. The nystagmus observed in two patients with Dr was in association with malformations of the nervous system. It is interesting to see that there were no cataracts, lens abnormalities, colobomas, or macular abnormalities found with Dq-. The significance of this is difficult to assess due to the small size of the sample, and the fact that they were an infrequent manifestation in Dr. There remains the high incidence of retinoblastoma in 13q- which has been discussed above.

In conclusion, it is interesting to compare the 13q- deletions and 13 ring chromosomes with trisomy 13. Chromosome 13 is acrocentric with most of its genetic material on the long arm so that theoretically, trisomy 13 and 13q- should produce a type and countertype. The features which the two have in common are microcephaly, malformed ears, micrognathia, and malformed feet. In trisomy 13, however, there is polydactyly, while in 13q- and 13r, absent thumbs are common. The ophthalmological abnormalities of trisomy 13 are well documented in nearly 200 cases (11), and the main features are microphthalmos, coloboma, cataract, and retinal dysplasia. In the cases of 13q- and 13r mentioned earlier, there are a few with microphthalmos and coloboma, but only one with a cataract (a Dr). And while retinal dysplasia is a characteristic finding in trisomy 13, retinoblastoma may occur with 13q- (Figs. 5-3 and 5-4). Cagianut and Theiler (8) commented that deletions of group D chromosomes tended to produce an inhibition of the normal development of retina and choroid, whereas trisomies produced retinal dysplasia characterized by proliferation. It now appears possible that in one of the two dense bands on the long arm of chromosome 13 (either 13q21 or 13q31), there is an area responsible for normal retinal development. A gene mutation in this area will result in the autosomal dominant form of retinoblastoma. A deletion of this area will produce the 13q- syndrome of retinoblastoma and mental deficiency, while a trisomy results in the condition of abnormal retinal development known as retinal dysplasia.

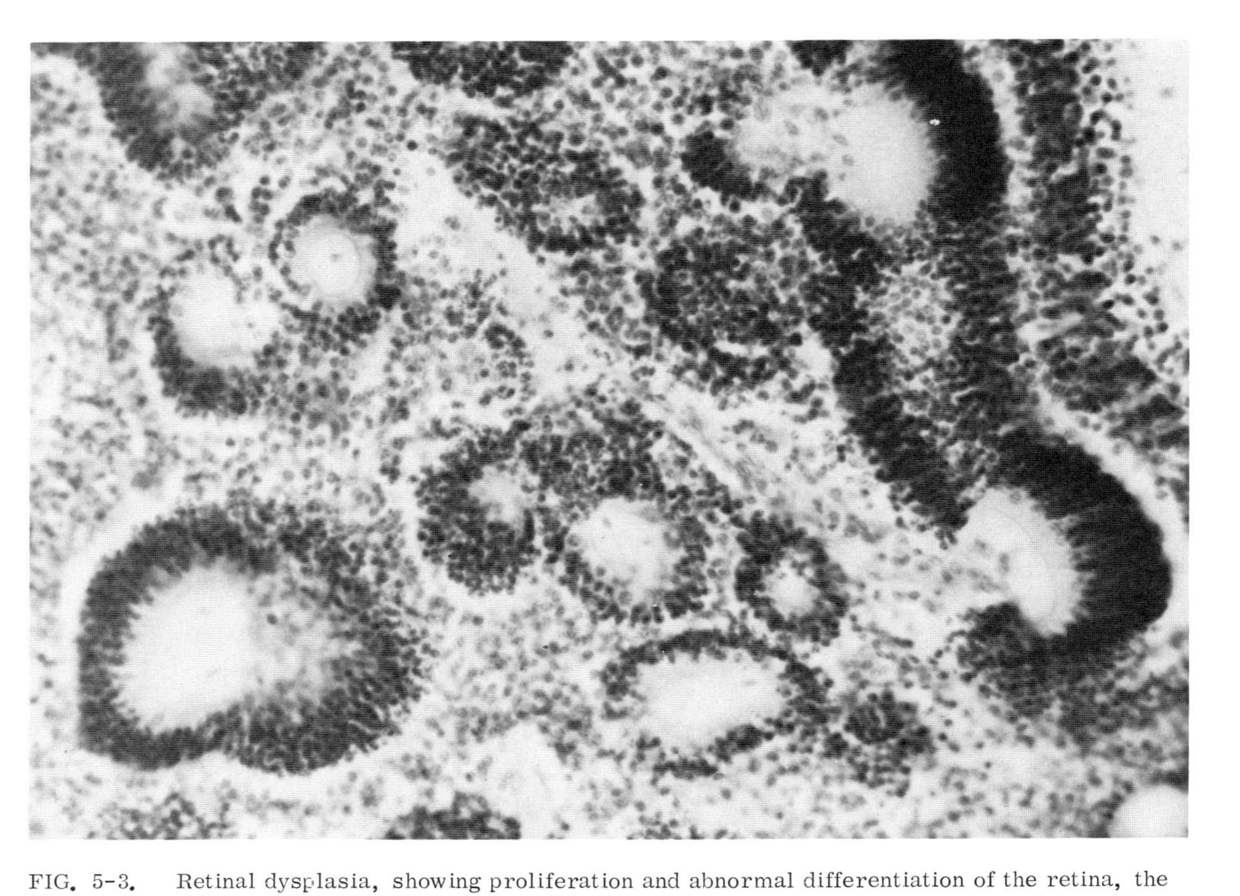

FIG. 5-3. Retinal dysplasia, showing proliferation and abnormal differentiation of the retina, the neuroepithelial cells being arranged in rosettes. (By permission of Professor Norman Ashton, Institute of Ophthalmology, London)

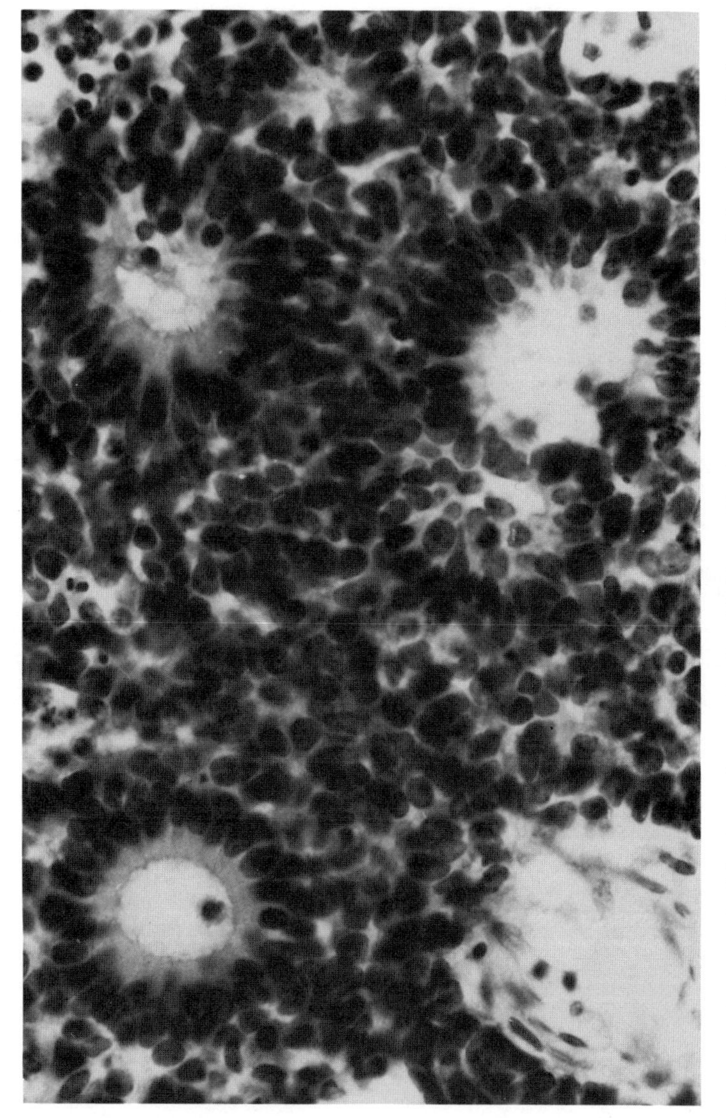

FIG. 5-4. Retinoblastoma in a patient with 13q−, the tumour consisting of closely packed undifferentiated and anaplastic cells and containing groups of cells arranged in rosettes. (By permission of Mr. A. J. Bron)

TABLE 5-2

Group D Chromosomes: Deletions Dq-

Reference	Ptosis	Antimongoloid Slant	Hypertelorism	Epicanthus	Microphthalmos	Strabismus	Iris Coloboma	Choroidal Coloboma	Retinoblastoma	Mental Retardation	Chromosome Abnormality	Remarks
Lele et al. (29) 1963									+	+	15q-	
Thompson and Lyons (52) 1965					+		+		(+)	+	Dq-	Complex mosaic with a Dq- cell line and a probable D/C translocation
de Grouchy et al. (17) 1966			+								Dp-	Almost lateral disposition of orbits
van Kempen (23) 1966		+		+					+	+	Dq-	Dq- and t(2:C)
Mikelsaar (31) 1967			+	+							Dq-	Chromosome mosaic
Allderdice et al. (2) 1969 Case 1	+	+	+		+		+				Dq-	Small chromosomal deletion

TABLE 5-2 (continued)

Reference	Ptosis	Antimongoloid Slant	Hypertelorism	Epicanthus	Microphthalmos	Strabismus	Iris Coloboma	Choroidal Coloboma	Retinoblastoma	Mental Retardation	Chromosome Abnormality	Remarks
Opitz et al. (37) 1969							+	+			Dq-	Asymmetry of palpebral fissures and of the corneae
Parker et al. (41) 1969		+				Con					13p-	
Wilson et al. (58) 1969									+	+	14q-	See subsequent paper, Wilson et al. (59) for banding studies on same patient
Cagianut and Theiler (8) 1970							+	+			14 or 15q-	Detailed description of ocular pathology; retina thrown up in folds, retarded development of ciliary body, etc.
Gey (14) 1970	+					Con			+		Dq-	LE: Megalocornea, proptosis and convergent strabismus
Orbeli and Luriye (38) 1970	+				+						Dq-	Proptosis; chromosome mosaic

Taylor (50) 1970 Case 1				+						Dq-	Mongoloid slant; corneal opacity RE
Case 2			+					+		Dq-	Chromosome mosaic
Laurent et al. (26) 1971		+			Div					15q-	Amaurotic paralysis of left pupil; pale papillae
Orbeli et al. (39) 1971	+			+						Dq-	Proptosis; chromosome mosaic
Orye et al. (40) 1971					Con			+	+	13q-	
Wilson et al. (59) 1973								+		13q-	Description of banding studies on patient previously described by authors (58)
Howard et al. (20) 1974			+					+	+	13q-	Banding studies; long eyelashes; prominent eyebrows
O'Grady et al. (36) 1974 Case 1						(+)	(+)	(+)	+	Dq-	Retinoblastoma LE; colobomas RE
Case 2								(+)		Dq-	

TABLE 5-3

Comparison of Ocular Findings in Partial Monosomy D

Ocular Abnormality	Dq-,* 18 Cases	Dr, 34 Cases
Ptosis	4	2
Hypertelorism	5	19
Antimongoloid slant	3	2
Mongoloid slant	1	5
Narrow palpebral fissures	0	4
Epicanthus	2	21
Microphthalmos	5	6
Strabismus	3	5
Nystagmus	0	2
Iris coloboma	5	4
Uveal coloboma	1	2
Cataract	0	2
Lens coloboma	0	2
Macular abnormality	?1	3
Retinoblastoma	10	1

* The two cases of Dp- are not included in the totals.

REFERENCES

1. Adams, M. S. Palm-prints and a ring-D chromosome. Lancet, 2: 494-495 (1965).

2. Allderdice, P. W., Davis, J. G., Miller, O. J., Klinger, H. P., Warburton, D., Miller, D. A., Allen, F. H. Jr., Abrams, C. A. L. and McGilvray, E. The 13q deletion syndrome. Am. J. Hum. Genet. 21: 499-512 (1969).

3. Ayraud, N. and Szepetowski, G. Un cas de chromosome D en anneau. Ann. Genet. (Paris) 12: 259-261 (1969).

4. Bain, A. D. and Gauld, I. K. Multiple congenital abnormalities associated with ring chromosome. Lancet 2: 304-305 (1963).

5. Bilchik, R. C., Zackai, E. H., Smith, M. E. and Williams, J. D. Anomalies with ring D chromosome. Am. J. Ophthalmol. 73: 83-89 (1972).

6. Biles, A. R. Jr., Luers, Th. and Sperling, K. D ring chromosome in newborn with peculiar face, polydactyly, imperforate anus, arrhinencephaly, and other malformations. J. Med. Genet. 7: 399-401 (1970).

7. Bloom, G. E., Gerald, P. S. and Reisman, L. E. Ring D chromosome: A second case associated with anomalous haptoglobin inheritance. Science 156: 1746-1748 (1967).

8. Cagianut, B. and Theiler, K. Bilateral colobomas of iris and choroid. Association with partial deletion of a chromosome of group D. Arch. Ophthalmol. 83: 141-144 (1970).

9. Coffin, C. S. and Wilson, M. G. Ring chromosome D (13). Am. J. Dis. Child. 119: 370-373 (1970).

10. Day, R. W., Wright, S. W., Koons, A. and Quigley, M. XXX 21-trisomy and retinoblastoma. Lancet 2: 154-155 (1963).

11. François, J., Berger, R. and Saraux, H. Les Aberrations Chromosomiques en Ophtalmologie. Paris, Masson et Cie, 1972, pp.231-258.

12. Fryns, J. P., Deroover, J., van den Berghe, H., Cassiman, J. J., Goffaux, P. and Lebas, E. Malformative syndrome with ring chromosome 13. Humangenetik 24: 235-240 (1974).

13. Gerald, P. S., Warner, S., Singer, J. D., Corcoran, P. A. and Umansky, I. A ring D chromosome and anomalous inheritance of haptoglobin type. J. Pediatr. 70: 172-179 (1967).

14. Gey, W. Dq-, multiple Missbildungen und Retinoblastom. Humangenetik 10: 362-365 (1970).

15. Gilgenkrantz, S., Cabrol, C., Lausecker, C., Hartley, B. M. E. and Bohe, B. Le syndrome Dr. Etude d'un nouveau cas (46, XX, 14r). Ann. Genet. (Paris) 14: 23-31 (1971).

16. Grace, E., Drennan, J., Colver, D. and Gordon, R. R. The 13q- deletion syndrome. J. Med. Genet. 8: 351-357 (1971).

17. Grouchy, J. de, Salmon, Ch., Salmon, D. and Maroteaux, P. Délétion du bras court d'un chromosome 13-15, hypertélorisme et phénotype haptoglobine Hp0 dans une même famille. Ann. Genet. (Paris) 9: 80-85 (1966).

18. Hollowell, J. G., Littlefield, L. G., Dharmkrong-At, A., Folger, G. M., Heath, C. W., jun. and Bloom, G. E. Ring 13 chromosome with normal haptoglobin inheritance. J. Med. Genet. 8: 222-226 (1971).

19. Hoo, J. J., Obermann, U. and Cramer, H. The behavior of ring chromosome 13. Humangenetik 24: 161-171 (1974).

20. Howard, R. O., Breg, W. R., Albert, D. M. and Lesser, R. L. Retinoblastoma and chromosome abnormality. Arch. Ophthalmol. 92: 490-493 (1974).

21. Jensen, R. D. and Miller, R. W. Retinoblastoma: Epidemiologic characteristics. N. Engl. J. Med. 285: 307-311 (1971).

22. Juberg, R. C., Adams, M. S., Venema, W. J. and Hart, M. G. Multiple congenital anomalies associated with a ring-D chromosome. J. Med. Genet. 66: 314-321 (1969).

23. Kempen, van C. A case of retinoblastoma combined with severe mental retardation and a few other anomalies associated with complex aberrations of the caryotype. Maandschr. Kindergeneeskd. 34: 92-95 (1966).

24. Kistenmacher, M. L. and Punnet, H. H. Comparative behaviour of ring chromosomes. Am. J. Hum. Genet. 22: 304-318 (1970).

25. Ladda, R., Atkins, L., Littlefield, J. and Pruett, R. Retinoblastoma: Chromosome banding in patients with heritable tumour. Lancet 2: 506 (1973).

26. Laurent, C., Noël, B. and David, M. Essai de classification des délétions du bras long d'un chromosome du groupe D. A propos d'un case 15q-. Ann. Genet. (Paris) 14: 33-40 (1971).

27. Lehrke, R., Thelen, T. and Lehre, R. Jr. Syndrome associated with group D chromosome deletions. Lancet 2: 98-99 (1971).

28. Lejeune, J., Lafourcade, J., Berger, R., Cruveiller, J., Réthoré, M. -O., Dutrillaux, B., Abonyi, D. and Jérôme, H. Le phénotype Dr. Etude de 3 cas de chromosomes D en anneau. Ann. Genet. (Paris) 11: 79-87 (1968).

29. Lele, K. P., Penrose, L. S. and Stallard, H. B. Chromosome deletion in a case of retinoblastoma. Ann. Hum. Genet. 27: 171-174 (1963).

30. Masterson, J. G., Law, E. M., Rashad, M. N., Cahalane, S. F. and Kavanagh, T. M. A malformation syndrome with ring D chromosome. J. Ir. Med. Assoc. 61: 398-399 (1968).

31. Mikelsaar, A. -V. N. The mosaicity with respect to the deletion of a part of the long arm of one of the chromosomes of group D (in man). Genetika 3(4): 142-145 (1967).

32. Mikkelsen, M. and Niebuhr, E. A ring chromosome (46,XY,13r) occurring in a family with a D-D translocation 13-,14-,t(13q14q). Ann. Genet. (Paris) 12: 51-56 (1969).

33. Morič-Petrovič, S., Garzicic, B., Despotovic, M. and Kaličanin, P. Une observation de chromosome D en anneau. Ann. Genet. (Paris) 13: 265-267 (1970).

34. Neimann, N., Ducas, J., Gilgenkrantz, S. and Peters, A. Un cas de chromosome en anneau du groupe 13-15. Arch. Fr. Pediatr. 24: 584 (1967).

35. Niebuhr, E. and Ottosen, J. Ring chromosome D (13) associated with multiple congenital malformations. Ann. Genet. (Paris) 16: 157-166 (1973).

36. O'Grady, R. B., Rothstein, T. B. and Romano, P. E. D- group deletion syndromes and retinoblastoma. Am. J. Ophthalmol. 77: 40-45 (1974).

37. Opitz, J. M., Slungaard, R., Edwards, R. H., Inhorn, S. L., Muller, J. and Venecia, G. de. Report of a patient with a presumed Dq- syndrome. Birth Defects: Orig. Art. Ser. 5(pt 5): 93-99 (1969).

38. Orbeli, D. I. and Luriye, I. V. A syndrome associated with a deletion in the long arm of a D group chromosome. Genetika 6(12): 116 (1970).

39. Orbeli, D. J., Lurie, I. W. and Goroshenko, Ju. L. The syndrome associated with the partial D-monosomy. Humangenetik 13: 296-308 (1971).

40. Orye, E., Delbeke, M. J. and Vandenabeete, B. Retinoblastoma and D-chromosome deletions. Lancet 2: 1376 (1971).

41. Parker, C. E., Koch, R., Mavalwala, J., Derencsenyi, A. and Hatashita, A. Familial deletion of the short arm of the D_1 chromosome (46,XX,13p-) not associated with loss of haptoglobin or catalase activity. Clin. Pediatr. (Phila.) 8: 453-458 (1969).

42. Picciano, D. J., Berlin, C. M., Davenport, S. L. H. and Jacobson, C. B. Human ring chromosomes: A report of five cases. Ann. Genet. (Paris) 15: 241-247 (1972).

43. Pruett, R. C. and Atkins, L. Chromosome studies in patients with retinoblastoma. Arch. Ophthalmol. 82: 177-181 (1969).

44. Réthoré, M. -O., Praud, E., Le Loch, J., Joly, C., Saraux, H., Aussannaire, M. and Lejeune, J. Diagnostic clinique de phénotype correspondant à un chromosome D en anneau. Presse Med. 78: 955-958 (1970).

45. Réthoré, M. -O., Saraux, H., Prieur, M., Dutrillaux, B., Meer, J. -J. and Lejeune, J. Syndrome 48,XXY,+21 et rétinoblastome. Arch. Fr. Pediatr. 29: 533-538 (1972).

46. Salamanca, F., Buentello, L. and Armendares, S. Ring D1 chromosome with remarkable morphological variation in a boy with mental retardation. Ann. Genet. (Paris) 15: 183-186 (1972).

47. Saraux, H. Types et contre-types en pathologie chromosomique. Bull. Mem. Soc. Fr. Ophtalmol. 85: 8-16 (1972).

48. Saraux, H., Réthoré, M. -O., Aussannaire, M., Dhermy, P., Joly, C., Le Loch, J., Praud, E. and Lejeune, J. Les anomalies oculaires du phénotype DR (chromosome D en anneau). Ann. Ocul. (Paris) 203: 737-748 (1970).

49. Sparkes, R. S., Carrel, R. E. and Wright, S. W. Absent thumbs with a ring D2 chromosome - A new deletion syndrome. Am. J. Hum. Genet. 19: 644-659 (1967).

50. Taylor, A. I. Dq-, Dr and retinoblastoma. Humangenetik 10: 209-217 (1970).

51. Teplitz, R. L., Miller, D., Hansson, K. M. and Rundell, T. S. A human ring D chromosome associated with multiple congenital abnormalities. J. Pediatr. 70: 936-941 (1967).

52. Thompson, H. and Lyons, R. B. Retinoblastoma and multiple congenital anomalies associated with complex mosaicism with deletion of D chromosome and probable D/C translocation. Human Chromosome Newsletter 15: 21 (1965).

53. Tolksdorf, M., Goll, U., Wiedemann, H. -R. and Pfeiffer, R. A. Die Symptomatik von Ringchromosomen der D-Gruppe. Arch. Kinderheiklkd. 181: 282-295 (1970).

54. Tolksdorf, M., Wiedmann, H. -R. and Goll, U. Ring D1 chromosome and multiple malformations. Lancet 2: 1009 (1969).

55. Varela, M. A. and Sternberg, W. H. Ring chromosomes in two infants with congenital malformations. J. Med. Genet. 6: 334-341 (1969).

56. Wang, H. C., Melnyk, J., McDonald, L. T., Uchida, I. A., Carr, D. H. and Goldberg, B. Ring chromosomes in human beings. Nature (Lond), 195: 733-734 (1962).

57. Wiener, S., Reese, A. B. and Hyman, G. A. Chromosome studies in retinoblastoma. Arch. Ophthalmol. 69: 311-313 (1963).

58. Wilson, M. G., Melnyk, J. and Towner, J. W. Retinoblastoma and deletion D(14) syndrome. J. Med. Genet. 6: 322-327 (1969).

59. Wilson, M. G., Towner, J. W. and Fujimoto, A. Retinoblastoma and D-chromosome deletions. Am. J. Hum. Genet. 25: 57-61 (1973).

CHAPTER 6

DEFICIENCIES OF GROUP E CHROMOSOMES

I. DELETION OF THE SHORT ARM OF CHROMOSOME 18 (18p-)

The syndrome of the deletion of the short arm of chromosome 18, 18p-, was first described by de Grouchy et al. (44), and at that time the deletion was thought to be of chromosome 17 or 18. Since then there have been over 30 cases of 18p- associated with ocular abnormalities ranging in severity from hypertelorism to cyclopia.

A. Phenotype

In a recent review of 28 cases of 18p-, Fischer et al. (30) described the main features of the syndrome as: mental deficiency, a variety of eye defects, flat or saddle nose, malformed ears, dental caries, short stature, short neck with or without webbing and mild deformities of the fingers and toes (Fig. 6-1). The same authors considered that there are two forms of the syndrome because five of the 28 cases they reviewed had various grades of holoprosencephaly incompatible with life, and the other 23 viable cases had mild signs of craniofacial dysplasia. It is for this reason that holoprosencephaly has been included in the ocular abnormalities associated with 18p- which are summarized in Table 6-1. Holoprosencephaly is a group of abnormalities with anomalous development of the nose, mouth and ears, defective development of the olfactory and optic structures, and impaired midline cleavage of the embryonic forebrain. Arhinencephaly is the neurological part of this syndrome and cebocephaly is another severe form characterized by a single nostril and monoventricular arhinencephaly.

The first two references, those of de Grouchy et al. (44) and Thieffry et al. (101) refer to the same case which has also been referred to in the review by de Grouchy (41).

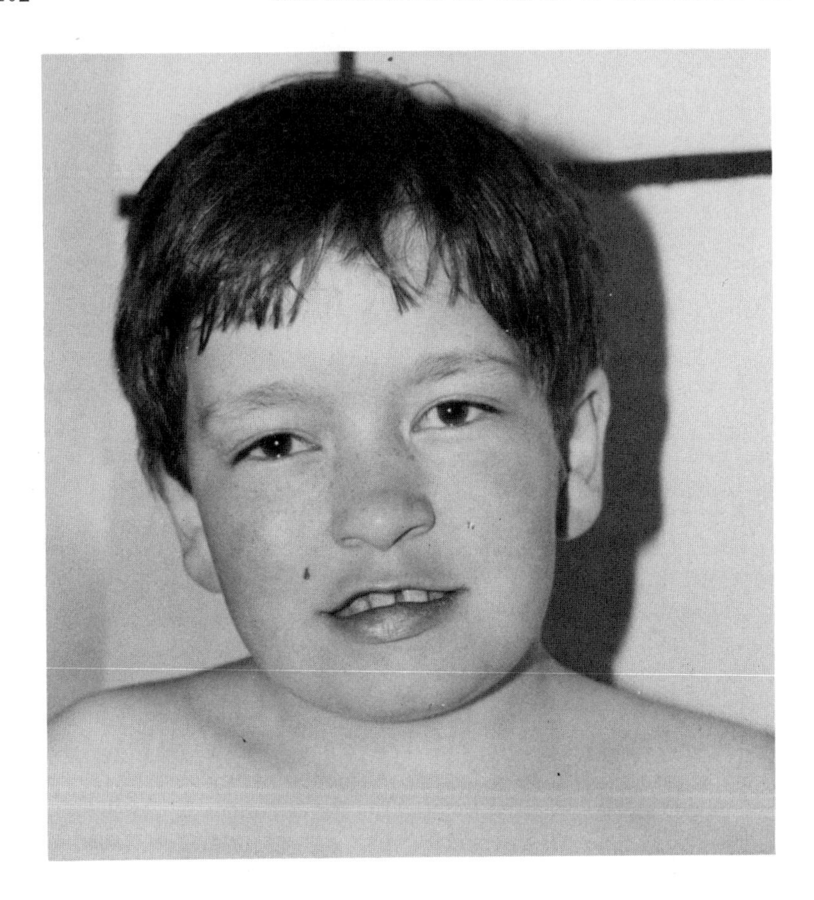

FIG. 6-1. 46, XX, 18p- with saddle nose and short neck. (By permission of Mr. L. J. Butler)

B. Ocular Abnormalities

The most common findings in a total of 37 cases are:

Hypertelorism, 17/37 or 46%

Epicanthus, 13/37 or 35%

Ptosis, 15/37 or 40%

Strabismus, 7/37 or 19%

These are relatively common abnormalities which are observed to a greater or lesser degree with every chromosomal abnormality and also with a normal karyotype. Abnormalities less commonly seen are nystagmus (two cases), antimongoloid slant (four cases), and mongoloid slant (two cases). It is interesting that there are no reports of iris colobomas or malformations of the anterior chamber as in the D group deletion syndromes. There are single examples of narrow palpebral fissures (101), deep corneal opacity and posterior keratoconus (73), myopia and astigmatism (30), amblyopia (62), and atrophic peripheral iris stroma (53). There is also a variety of abnormal fundi, with tapeto-retinal degeneration (53), brunescent posterior pole (67), and atypical pigmentation of the retina resembling vitiligo (71). These abnormalities are listed in Table 6-1.

Of interest to geneticists is the association of deficient or absent IgA with 18p-, first described by Ruvelcaba and Thuline (92), and subsequently by other investigators. Another aspect of this syndrome is the number of cases in which there was an actual or presumed translocation carrier or chromosomal mosaicism in one of the parents. Deletions arising out of translocations are quite common and the mechanism of this is described elsewhere.

The family described by Uchida et al. (103) has been widely quoted and consists of a mother and her two children by different fathers. The mother who was mentally retarded showed mosaicism for 18p- and both her children were affected, although one had the severe form of the syndrome with cebocephaly while the other was only mildly affected. The authors suggested that the severe form of the syndrome might be due to mutant recessive alleles that were present in a hemizygous state on the short arm of the normal chromosome 18. This hypothesis has been supported by Gorlin et al. (39) who also offered the alternative possibility of a homozygous state of an autosomal recessive gene in a person with a normal karyotype.

The five cases of holoprosencephaly illustrate the severe form of the syndrome and the two cases with cyclopia (28, 83) are of particular interest to ophthalmologists. Cyclopia is by no means always associated with a chromosomal abnormality and it has been reported in several instances with a normal karyotype. In addition, however, to the two cases mentioned above with 18p-, there has been one case with chromosomal mosaicism consisting of a monosomy G in 22 cells out of the 104 analysed (16), one case with an extra minute fragment (88), one with an 18r chromosome (18) and no fewer than eight cases with trisomy 13 (2, 5, 33, 38, 48, 100, 102). Even though the occurrence of ocular abnormalities seems to be the rule in trisomy 13, this number of cases of cyclopia is high particularly as cyclopia is a very rare abnormality. Cyclopia may be considered an extreme form of arhinencephaly and the hyopothesis of Fischer et al. (30) may explain the association of arhinencephaly with trisomy 13, since Gorlin et al. (39) have already accounted for the cases of arhinencephaly with normal karyotype. Fischer and coworkers (30) suggested that the presence of genes on the short arm of chromosome 18 controlled loci on chromosome 13, particularly the locus responsible for arhinencephaly. In a normal subject with a normal karyotype, one or other of the short arms of chromosome 18 will prevent the activation of loci on chromosome 13. In cases of trisomy 13, the excess of genetic material in chromosome 13 will produce arhinencephaly. In cases of a deletion of a short arm of chromosome 18, the form of the resulting syndrome will depend on the relevant preponderance of dominant and recessive alleles on the remaining short arm of chromosome 18. Another locus on chromosome 13, not associated with chromosome 18, would cause the other malformations associated with trisomy 13.

TABLE 6-1

Group E Chromosomes: 18p-

Reference	Arhinencephaly	Cyclopia	Ptosis	Antimongoloid Slant	Hypertelorism	Hypotelorism	Epicanthus	Microphthalmos	Strabismus	Nystagmus	Remarks
de Grouchy et al. (44) 1963					+				Con		
Thieffry et al. (101) 1963				+							Narrow palpebral fissures
Bühler et al. (10) 1964							+				
Faint and Lewis (28) 1964		+									Single globe, no optic nerve
Summitt (99) 1964											Slight hypoplastic supra-orbital ridges; pupils slightly eccentric
van Dyke et al. (26) 1964									Con		

TABLE 6-1 (continued)

Reference	Arhinencephaly	Cyclopia	Ptosis	Antimongoloid Slant	Hypertelorism	Hypotelorism	Epicanthus	Microphthalmos	Strabismus	Nystagmus	Remarks
Uchida et al. (103) 1965			+				+				Also a sister with cebocephaly and hypotelorism; maternal translocation
de Grouchy et al. (42) 1966			(+)								
Lejeune et al. (64) 1966b			+	+	+		+				Maternal 18p-;17p+
Migeon (73) 1966							+				Deep corneal opacity and posterior keratoconus of LE
Nitowski et al. (83) 1966		+									Fusion of 2 globes with 2 irides; pupils and corneae
Pfeiffer (87) 1966					+						Absent nasal bones
Reinwein et al. (90) 1967			+		+		+		Con		

Gilgenkrantz et al. (36) 1968		(+)	+		+		Div	+	Intermittent horizontal nystagmus
Gorlin et al. (39) 1968				+		+			Cebocephaly; pupils very small
Jacobsen and Mikkelsen (54) 1968b Case 1		+							Atrophic peripheral iris stroma; total alopecia
Case 2		+			+				Tapeto-retinal degeneration
Jacobsen and Mikkelsen (53) 1968a Case 1		+			+				Case 1 and 2: normal fundi; familial t(18p-;21p+)
Case 2			+		+				
McDermott et al. (69) 1968	+			+					The mother has the same deletion, but normal phenotype
Ayraud et al. (4) 1969			+						Fundi appear normal
Blumina and Podugolnikova (7) 1969			+		+				
Ruvelcaba and Thuline (92) 1969			+						Absent IgA
Schwanitz et al. (96) 1969		+	+						

TABLE 6-1 (continued)

Reference	Arhinencephaly	Cyclopia	Ptosis	Antimongoloid Slant	Hypertelorism	Hypotelorism	Epicanthus	Microphthalmos	Strabismus	Nystagmus	Remarks
Fischer et al. (80) 1970									Div		Myopia and astigmatism; normal fundi; absent IgA
Laurent et al. (59) 1970					+						Mongoloid slant; normal fundi; paternal ? 46, XY/46, XY, 18p-
Vaillaud et al. (104) 1970				+	+						
de Chieri et al. (14) 1971					+					+	Enophthalmos
Harlan et al. (49) 1971					+		+				
Levenson et al. (67) 1971			(+)								Brunescent posterior pole
Malpuech et al. (70) 1971a											Atypical pigmentation of the retina resembling vitiligo; t(G-;18p+) de novo; parents normal

Malpuech et al. (71) 1971b	+				Karyotype 45, XX, 18-, G-, t(18qGp)+
Morič-Petrovič et al. (79) 1971	+	+			
Cohen and Putnam (17) 1972	+		+		Karyotype 45, XX, 18-, 21-, t(18p21q)+; clinical findings compatible with 18p- syndrome
Gilgenkrantz et al. (35) 1972	+		+	+	Deficient IgA; enophthalmos t(22-;18p+); normal parents
Šubrt and Beránková (98) 1972			+	+	Slight mongoloid slant
Leisti et al. (62) 1973	(+)				Amblyopia; absence of IgA; karyotype 45, XY, 13-, 18-, t(13q18q)+

II. DELETIONS OF THE LONG ARM OF CHROMOSOME 18 (18q-)

The syndrome of 18q-, or deletion of the long arm of chromosome 18, was first described by de Grouchy et al. (46).

A. Phenotype

The main clinical features include mental retardation, short stature, microcephaly, midfacial dysplasia, and atretic ear canals. The mouth is often shaped like a carp, there are a variety of eye defects, and there are long tapering fingers with dimples at the knuckles. Dimples are also common at the shoulders, elbows, and knees. Congenital heart disease has been reported and also absent or deficient IgA, as in the 18p- syndrome. However, in contrast to the 18p- syndrome, there seems to be only one phenotype which appears to be fairly uniform. The subjects survive beyond infancy, since their abnormalities are not incompatible with life as they are in a few of the 18p- cases, and there are even a few instances of 18q- reported in an adult (27) (Figs. 6-2 and 6-3).

B. Ocular Abnormalities

The ocular abnormalities associated with 18q- are given in Table 6-2 and in a total of 40 cases those most commonly reported are hypertelorism (17 cases, 42.5%); epicanthus (13 cases, 32.5%); and strabismus (16 cases, 40%). These are the ocular abnormalities common to most chromosomal syndromes and it is only their absence which is of any significance. Enophthalmos is fairly common (4 cases, 10%) and may be associated with the nasomaxillary hypoplasia which occurs in this syndrome producing eyes that appear deep-set. Colobomas are unusual findings: an iris coloboma being reported by Gleissner et al. (37), a choroidal coloboma by Wertelecki and Gerald (108), and a coloboma of the iris, choroid and optic nerve by Mikkelsaar and Talvik (74, 75). Nystagmus is a characteristic feature of the syndrome occurring in 12 cases (30%), whereas it is unusual when found in the 18p- and 18r syndromes, and when found in the 18q- syndrome, it is not associated with neurological abnormalities. Another characteristic feature of this syndrome, which does not occur in the 18p- and 18r syndromes, is optic atrophy. There are five instances of optic atrophy including one possible case, and five cases of pale or very pale discs. For the most part, these cases were not examined by an ophthalmologist so that it is rather difficult to establish which can be considered as genuine optic atrophies. However, according to François et al. (32), a partial optic atrophy affecting the temporal sector was described by several groups of authors (24, 52, 109, 110). Wertelecki and Gerald (108) also described posterior staphylomas and temporal tilting of the optic discs in their observations, which included four cases previously described in 1966. Another finding is pale discs (1, 24, 77, 81), and the original case of de Grouchy et al. (46) had a tapeto-retinal degeneration.

Anomalies of the anterior segment are an unusual finding (60, 63) and may correspond to a cleavage syndrome. In addition, microcornea was seen by Law and Masterson (60, 61), myopia and astigmatism, as in the case described by Destiné et al. (24), or severe myopia as described by Wertelecki and Gerald (108), are the clinical features associated with pale discs or optic atrophy.

In most cases, the karyotype of the parents was normal but there are seven instances of a parental balanced translocation, and one case in which the father was a mosaic (21). In one of the six cases described by Wertelecki and Gerald (108), the father and siblings of the propositus had a pericentric inversion of chromosome 18.

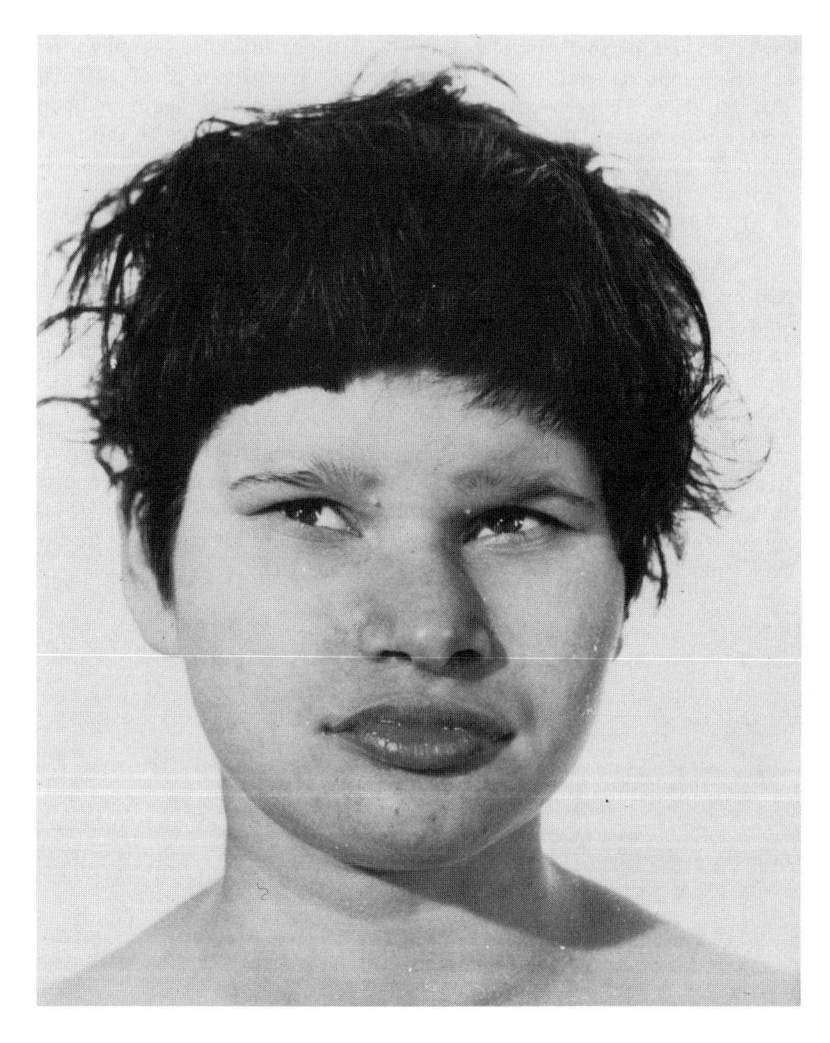

FIG. 6-2. 46,XX,18q- with abnormal facies and mongoloid slant.

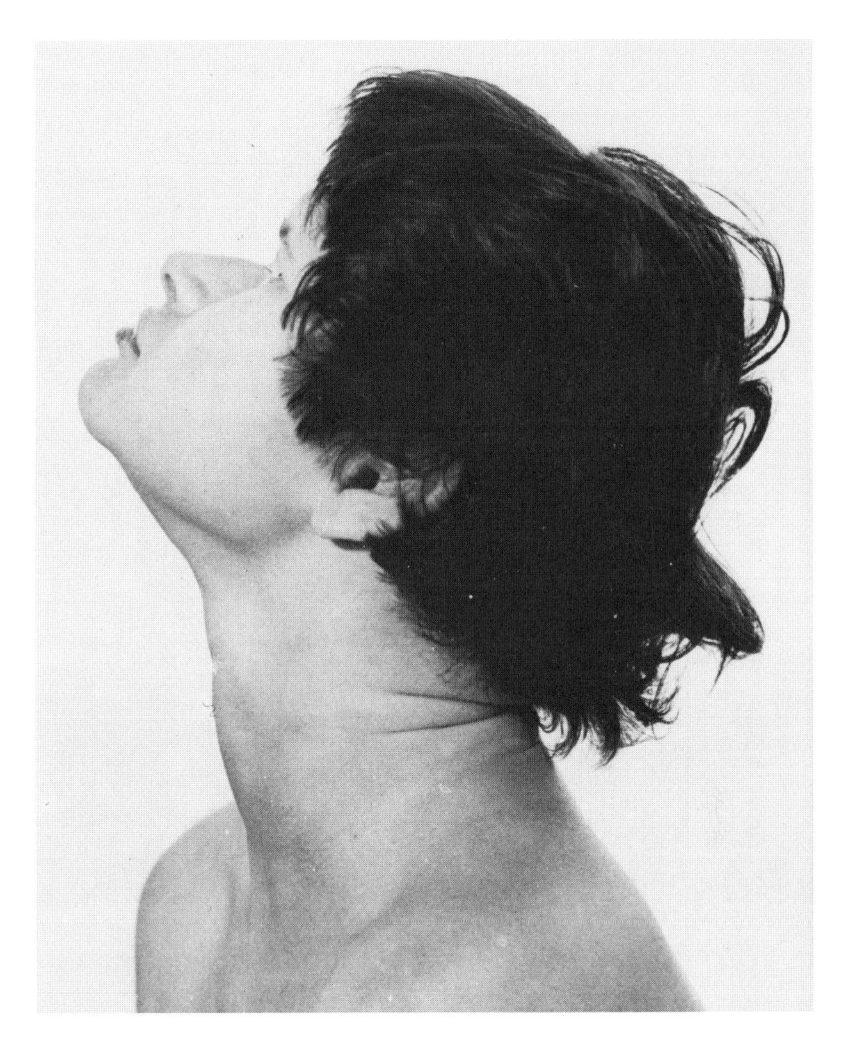

FIG. 6-3. Profile of Fig. 6-2, showing depressed bridge of the nose. (By permission of Dr. Renata Lax and Dr. Michael Ridler)

TABLE 6-2

Group E Chromosomes: 18q-

Reference	Enophthalmos	Mongoloid Slant	Antimongoloid Slant	Narrow Palpebral Fissures	Hypertelorism	Epicanthus	Strabismus	Nystagmus	Remarks
de Grouchy et al. (46) 1964b								+	Oscillatory nystagmus; tapeto-retinal degeneration
Law and Masterson (61) 1966						+			Microcornea
Lejeune et al. (63) 1966a Case 1	+							+	"Vision diminished" at 3 years of age; described previously by de Grouchy et al. (46)
Case 2	+		+		+				
Case 3								+	Bilateral corneal opacities; RE: absent anterior chamber; LE: microphthalmos, very shallow anterior chamber
Wertelecki et al. (109) 1966								3/4	Ocular fundoscopic anomalies 4/4

Day et al. (21) 1967							L Con		? Optic atrophy; paternal chromosome mosaicism
Destiné et al. (24) 1967 Case 1				+	+	+			Astigmatism and myopia; pale discs on temporal side
Case 2		+				+	Div	+	Bilateral optic atrophy; macular anomalies; horizontal nystagmus
Insley (52) 1967 Case 1						+	L Con		Bilateral optic atrophy
Case 2	+	+		+					
Valdmanis et al. (105) 1967 Case III-2			+			+			Oval pupils; artery in medial aspect of upper canthus of sclera; paternal t(18q-;4p+)
Wolf et al. (110) 1967					+	+	Con		"Ocular fundoscopic anomalies"
Nance et al. (81) 1968						+	L Con	+	Horizontal nystagmus; pale optic discs
Aarskog (1) 1969					+	+			Very pale optic discs; maternal t(18p-;3?+) and t(18q-;3?+); clinical features of 18q-
Borkowf et al. (8) 1969				+	+		Con		
Cenani et al. (12) 1969b					+	+			? Optic atrophy
Feingold et al. (29) 1969 Case 2							Con		Normal fundi

TABLE 6-2 (continued)

Reference	Enophthalmos	Mongoloid Slant	Antimongoloid Slant	Narrow Palpebral Fissures	Hypertelorism	Epicanthus	Strabismus	Nystagmus	Remarks
Law and Masterson (60) 1969 Case 1									Small corneae; brother of Case 2
Case 2		+				+			RE: microphthalmos, microcornea, very shallow anterior chamber, diffuse atrophy of the iris, posterior opacity of the lens
Lejeune et al. (65) 1969								+	Telecanthus
Mikkelsaar and Talvik (74) 1969 (75) 1970							Con		LE: coloboma of the iris, choroid and optic nerve
Sanroman and Rethore (93) 1969					+				Brushfield spots
Curran et al. (20) 1970					+				

Gleissner et al. (37) 1970			+				Iris coloboma; maternal translocation
Moore and Engel (77) 1970a				+	L Con	+	Pale optic discs; horizontal nystagmus
Stewart et al. (97) 1970 Case 1			+		+	+	Absent IgA
Fraccaro et al. (31) 1971			+				Ptosis of right lid
Hoehn et al. (50) 1971		+	+		+		Paternal t(4q-;18q+)
Jacobsen et al. (55) 1971	+						Blue sclerae; excessive myopia; bilateral optic atrophy; paternal t(18q-;6p+)
Wertelecki and Gerald (108) 1971 Case 1							Eccentric pupils; temporally tilted discs; posterior staphyloma
Case 2		+	+		+		Severe myopia
Case 3		+	+				Oval cornea; iris attachment to cornea; chor- oidal colobomata; temporally tilted discs
Case 4							No ocular abnormalities
Case 5					Con		
Case 6			+				Temporally tilted discs within a staphyloma; pupillary remnants; paternal inv(18p+q-)

TABLE 6-2 (continued)

Reference	Enophthalmos	Mongoloid Slant	Antimongoloid Slant	Narrow Palpebral Fissures	Hypertelorism	Epicanthus	Strabismus	Nystagmus	Remarks
Faed et al. (27) 1972b						+	Alt		
Gouw et al. (40) 1973				+		+			
Schinzel et al. (95) 1975 Case 1					+		+	+	Case 1 and 2: de novo 18q-
Case 2			+						
Case 3			+		+		+	+	Iris and choroid coloboma; paternal t(15;18)(q24;q21); pale discs

III. RING CHROMOSOME 18 (18r)

Wang et al. (107) first described the behaviour of ring chromosomes in man, and one of their two cases was that of a mentally retarded child with hypertelorism and epicanthus, as well as other abnormalities, and a missing group E chromosome, probably chromosome 18, being replaced by a ring.

A. Phenotype

The major clinical features in the 18r syndrome include low birth weight, microcephaly, mental retardation, low-set ears, atresia of the external auditory meatus, abnormalities of the hands and feet and of the genito-urinary and cardiac systems.

B. Ocular Abnormalities

The ocular abnormalities associated with the 18r syndrome are summarized in Table 6-3, to which the one case of 16r (85) has been added for comparison. It will be seen that there is the customary high incidence of hypertelorism and epicanthus (50%) and strabismus (30.6%). Nystagmus, which is common in the 18q- syndrome, is seen in only three cases. Colobomas involving the iris, choroid and sometimes the retina are fairly common (22% of all cases).

There is only one detailed description of the ocular pathology in the 18r syndrome by Yanoff and his colleagues (111). Their patient had bilateral microphthalmos with a colobomatous cyst, an immature angle of the anterior chamber, hypoplasia of the iris, persistent tunica vasculosa lentis, cataract, coloboma of the choroid and sclera and of the retinal and ciliary body pigment epithelium, retinal dysplasia and nonattachment, and intraocular smooth muscle and cartilage. This description is typical of the findings in trisomy 13 and could be explained by the hypothesis of Fischer et al. (30) which establishes a connection between chromosomes 13 and 18. Another suggested explanation is that the genetic content of the chromosomal fragment, which is in excess or absent, does not directly influence the phenotype but produces its effect by influencing a particular stage in the development of the embryo.

The formation of ring chromosomes has been briefly discussed previously and it would not be surprising to find similar abnormalities in the 18r, 18p- and 18q- syndromes. The ocular findings in the partial monosomies 18 are compared in Table 6-4 in which the signs common to each syndrome are apparent. It is interesting that holoprosencephaly occurs in the 18r (four cases, including one with cyclopia (18)) and in the 18p- (five cases), but not in the 18q- syndrome. In contrast, colobomas occur in the 18r (seven cases) and in the 18q- (four cases) but not in the 18p- syndrome. A similar comparison of all the clinical signs of the three syndromes is given in the excellent review by de Grouchy (41) and more recently by Kunze et al. (57) and Lurie and Lazjuik (68). Lurie and Lazjuik have also listed the incidences of the ocular abnormalities in the partial monosomies 18.

Finally, there is the interesting theory of type and countertype of trisomy 18 described by Saraux (94) in which he compares the general and, in particular, the ocular abnormalities occurring in trisomy 18 and in the deletion syndromes of chromosome 18. Theoretically, if an abnormality is present in a trisomy 18, it should be absent, or else its exact counterpart should occur in one of the deletion syndromes. This is illustrated by the colobomas of the optic nerve and the corneal ulcerations in trisomy 18, and the optic atrophy and anterior chamber cleavage syndromes seen in 18q- and 18r. Saraux also discusses the characteristic exaggerated development of the optic disc in trisomy 18, whereas the disc is small and often crescent-shaped in the 18 deletion syndromes. Trisomy 18 also produces abnormalities of the anterior part of the cornea, while it is the posterior part of the cornea which is affected in the deletion syndromes. These arguments are not conclusive; for example, there are not many cases of anterior chamber cleavage syndrome described in the deletion syndromes. It would seem, however, from all the literature on the 18 deletion syndromes, that Saraux's conclusions are well supported, namely that chromosome 18 influences the development of the optic nerve and of the cornea, whereas chromosome 13 influences the development of the intraocular structures and, in particular, the retina.

TABLE 6-3

Group E Chromosomes: 18r

Reference	Arhinencephaly	Cyclopia	Ptosis	Antimongoloid Slant	Mongoloid Slant	Hypertelorism	Epicanthus	Strabismus	Nystagmus	Iris Coloboma	Remarks
Wang et al. (107) 1962 Case 2						+	+				
Genest et al. (34) 1963						+	+				
de Grouchy et al. (45) 1964a						+	+				
Bernard et al. (6) 1966						+	+	Con			Microphthalmos
Lejeune et al. (66) 1966c						+					Brushfield spots
Aula et al. (3) 1967 Case 1						+					Myopia; bilateral corneal leucoma
Case 2											Retinae show numerous clear white spots

TABLE 6-3 (continued)

Reference	Arhinencephaly	Cyclopia	Ptosis	Antimongoloid Slant	Mongoloid Slant	Hypertelorism	Epicanthus	Strabismus	Nystagmus	Iris Coloboma	Remarks
Haenthens (47) 1967					+	+	+	Con			Normal fundi
Jeune et al. (56) 1967			(+)			+					
Mikkelsaar et al. (76) 1967						+				+	Bilateral colobomas of optic nerve and choroid
Palmer et al. (84) 1967				+		+		Div	+		
Petit and Poncelet (86) 1967			+						+	+	Pendular nystagmus; bilateral colobomas of optic nerve and choroid
de Grouchy et al. (43) 1968							+				
Hooft et al. (51) 1968					+	+	+	Con			

Brihaye et al. (9) 1969			+					+	+	Pendular nystagmus
Cenani et al. (11) 1969a					+	+	Con			
Deminatti et al. (23) 1969					+	+	Div			Small palpebral fissures; normal fundi
Feingold et al. (29) 1969 Case 1		(+)				+			+	Microphthalmos; bilateral uveal colobomas
Wald et al. (106) 1969		+							(+)	RE: choroidal coloboma; vessels of the fundi slightly attenuated
Christensen et al. (15) 1970 Mother				+			Con			Both cases: hypermetropia and astigmatism
Daughter							Con			
Deminatti et al. (2) 1970 Case 1					+	+	Div			
Case 2				+	+	+	Div			
Dumars et al. (25) 1970	+									Hypotelorism
Moore and Engel (78) 1970b Case 9									+	Right choroidal coloboma; 46, XX, 18r/45, XX, 18–
Murken et al. (80) 1970		+								

TABLE 6-3 (continued)

Reference	Arhinencephaly	Cyclopia	Ptosis	Antimongoloid Slant	Mongoloid Slant	Hypertelorism	Epicanthus	Strabismus	Nystagmus	Iris Coloboma	Remarks
Pergament et al. (85) 1970						+	+				Ring 16 chromosome
Richards et al. (91) 1970							+				Anisocoria
Yanoff et al. (111) 1970							+			+	Cebocephaly; cataract etc.; see detailed ocular pathology in text
Chang et al. (13) 1971											Absence of left optic nerve and eyeball
Cortesi and Patriarca (19) 1971						+	+	Con			
Michaels et al. (72) 1971					+		+				
Neu et al. (82) 1971											Bilateral absence of irides in lower nasal aspect; cebocephaly

Cohen et al. (18) 1972	+		
Kunze et al. (57) 1972		+	+
Picciano et al. (89) 1972		+	+

TABLE 6-4

Comparison of Ocular Findings in Partial Monosomy E

Ocular Abnormality	18p- (37 Cases)	18q- (40 Cases)	18r (35 Cases)
Hypertelorism	17	17	18
Narrow palpebral fissures	1	4	0
Ptosis	15	1	4
Epicanthus	13	13	8
Antimongoloid slant	4	4	2
Mongoloid slant	2	6	5
Enophthalmos	2	4	0
Microphthalmos	0	2	2
Myopia and astigmatism	1	3	2
Strabismus	7	16	11
Nystagmus	2	12	3
Glaucoma	0	1	0
Microcornea	0	2	0
Corneal opacities	1	1	1
Posterior keratoconus	1	0	0
Brushfield spots	0	1	1
Cataract	0	0	1
Coloboma iris/ciliary body	0	4	7
Macular abnormality	0	1	0
Optic atrophy	0	5 (+?5)	0
Cyclopia	2	0	1

REFERENCES

1. Aarskog, D. A familial 3/18 reciprocal translocation resulting in chromosome duplication deficiency (3?+;18q-). Acta Paediatr. Scand. 58: 397-406 (1969).

2. Arakaki, D. T. and Waxman, S. H. Trisomy D in a cyclops. J. Pediatr. 74: 620-622 (1969).

3. Aula, P., Gripenberg, U., Hjelt, L., Kivalo, E., Leisti, J., Palo, J., Schoultz, B. von and Suomalainen, E. Two cases with a ring chromosome in group E. Acta Neurol. Scand. 31(Suppl.): 51-52 (1967).

4. Ayraud, N., Darcourt, G., D'Oelsnitz, M., Poujol, J., Lavagna, J. and Capdeville, C. Syndrome 18p- . Une nouvelle observation. Ann. Genet. (Paris) 12: 122-125 (1969).

5. Batts, J. A. Jr., Punnett, H. H., Valdes-Dupena, M., Coles, J. W. and Green, W. R. A case of cyclopia. Am. J. Obstet. Gynecol. 112: 657-661 (1972).

6. Bernard, R., Stahl, A., Giraud, F., Hartung, M. and Brusquet, Y. Encéphalopathie avec dysmorphie complexe et chromosome 17-18 en anneau. Ann. Pediatr. (Paris) 42: 525-529 (1966).

7. Blumina, M. G. and Podugolnikova, O. A. Deletion of the short arm of the chromosome 18 in an infant with a peculiar form of oligophreny. Genetika 5(2): 148-157 (1969).

8. Borkowf, S. P., Wadia, R. P., Borgaonkar, D. S. and Bias, W. B. Partial deletion of the long arm of a chromosome 18. Birth Defects: Orig. Art. Ser. 5(pt. 5): 155-157 (1969).

9. Brihaye, M., Farnir, A., Poncelet, M. and Petit, P. Colobome irido-chorio-rétinien bilateral dans un cas de chromosome 18 en anneau. Bull. Soc. Belge Ophtalmol. 153: 696-701 (1969).

10. Bühler, E. M., Bühler, U. K. and Stalder, G. R. Partial monosomy 18 and anomaly of thyroxine synthesis. Lancet 1: 170 (1964).

11. Cenani, A., Pfeiffer, R. A. and Simon, H. A. Ring chromosome 18 (46, XX, 18r). Humangenetik 7: 351-352 (1969a).

12. Cenani, A., Schoeller, L. and Schubart, G. Deletion am langen Arm eines Chromosomes Nr 18 (46, XX, 18q-). Arch. Kinderheilkd. 178: 266-272 (1969b).

13. Chang, P., Perciaccante, R., Miller, O. J. and Rottino, A. Anophthalmia and other anomalies associated with a ring chromosome No 17-18. In Proceedings of the Fourth International Congress of Human Genetics, J. de Grouchy, F. J. C. Ebling, I. Henderson and J. François (Eds.), Amsterdam Excerpta Medica (International Congress Series No. 233), 1971, p. 44.

14. Chieri, P. R. de, Dedrato, A. and Albores, J. M. Possible 46,XX,18q-,18p- syndrome. J. Genet. Hum. (Paris) 19: 127-136 (1971).

15. Christensen, K. R., Friedrich, U., Jacobsen, P., Jensen, K., Nielsen, J. and Tsuboi, T. Ring chromosome 18 in mother and daughter. J. Ment. Defic. Res., 14: 49-67 (1970).

16. Cohen, M. M. Chromosomal mosaicism associated with a case of cyclopia. J. Pediatr. 69: 793-798 (1966).

17. Cohen, M. M. and Putnam, T. I. An 18p21q translocation in a patient with presumptive "monosomy G". Am. J. Dis. Child. 124: 908-910 (1972).

18. Cohen, M. M., Storm, D. F. and Capraro, V. J. A ring chromosome (no 18) in a cyclops. Clin. Genet. 3: 249-252 (1972).

19. Cortesi, M. and Patriarca, P. L. Ring type chromosome 18 [Chromosoma 18 ad annello - "deserizione di un caso e rassegna della letteratura"]. Minerva Pediatr. 23: 902-908 (1971).

20. Curran, J. P., Al-Salihi, F. L. and Allderdice, P. W. Partial deletion of the long arm of chromosome E-18. Pediatrics 46: 721-729 (1970).

21. Day, E. J., Marshall, R., MacDonald, P. A. C. and Davidson, W. M. Deleted chromosome 18 with paternal mosaicism. Lancet 2: 1307 (1967).

22. Deminatti, M., Debeugny, P., Croquette-Bulteel, M. F. and Delmas-Marsalet, Y. Deux observations de chromosome 18 en anneau. Ann. Genet. (Paris) 13: 149-155 (1970).

23. Deminatti, M., Dupuis, C., Maillard, E., Delmas-Marsalet, Y. and Bulteel, M. F. Une nouvelle observation de chromosome 18 en anneau. Ann. Genet. (Paris) 12: 126-129 (1969).

24. Destiné, M. L., Punnett, H. H., Thovichit, S., DiGeorge, A. M. and Weiss, L. La délétion partielle du bras long du chromosome 18 (syndrome 18q-). Rapport de deux cas. Ann. Genet. (Paris), 10: 65-69 (1967).

25. Dumars, K. W., Carnahan, L. G. and Barrett, R. V. Median facial cleft associated with ring E chromosome. J. Med. Genet. 7: 86-90 (1970).

26. Dyke, H. E. van, Valdmanis, A. and Mann, J. D. Probable deletion of the short arm of chromosome 18. Am. J. Hum. Genet. 16: 364-374 (1964).

27. Faed, M. J. W., Whyte, R., Paterson, C. R., McCathie, M. and Robertson, J. Deletion of the long arms of chromosome 18 (46,XX, 18q-) associated with absence of IgA and hypothyroidism in an adult. J. Med. Genet. 9: 102-104 (1972).

28. Faint, S. and Lewis, F. J. W. Presumptive deletion of the short arm of chromosome 18 in a cyclops. Human Chromosome Newsletter 14: 5 (1964).

29. Feingold, M., Schwartz, R. S., Atkins, L., Anderson, R., Bartsocas, C., Page, D. and Littlefield, J. W. IgA deficiency associated with partial deletion of chromosome 18. Am. J. Dis. Child. 117: 129-136 (1969).

30. Fischer, P., Golob, E., Friedrich, F., Kunze-Mühl, E., Doleschel, W. and Aichmair, H. Autosomal deletion syndrome, 46,XX,18p- : A new case report with absence of IgA in serum. J. Med. Genet. 7: 91-98 (1970).

31. Fraccaro, M., Hulten, M., Ivemark, B. I., Lindsten, J., Tiepolo, L. and Zetterqvist, P. Structural abnormalities of chromosome 18. I. A case of 18q- with autopsy findings. Ann. Genet. (Paris) 14: 275-280 (1971).

32. François, J., Berger, R. and Saraux, H. Les Aberrations Chromosomiques en Ophtalmologie. Paris, Masson et Cie, 1972, pp. 285-319.

33. Fujimoto, A., Ebbin, A. J., Towner, J. W. and Wilson, M. G. Trisomy 13 in two infants with cyclops. J. Med. Genet. 10: 294-304 (1973).

34. Genest, P., Leclerc, R. and Auger, C. Ring chromsome and partial translocation in the same cell. Lancet 1: 1426-1427 (1963).

35. Gilgenkrantz, S., Charles, J. -M., Cabrol, C., Mauuary, G. and Vigneron, C. Délétion du bras court du 18 par translocation t(22-:18p+) avec déficit en IgA. Ann. Genet. (Paris) 15: 275-281 (1972).

36. Gilgenkrantz, S., Marchal, C. and Neimann, N. La délétion du bras court du chromosome 18 (syndrome 18p-). A propos d'une nouvelle observation. Ann. Genet. (Paris) 11: 17-21 (1968).

37. Gleissner, M., Schwanitz, G. and Rott, H. D. Partielle Trisomie E18 (E18q-) als folge einer balancierten Translokation D/E bei der Mutter. Monatsschr. Kinderheilkd. 118: 441-444 (1970).

38. Golob, E., Schaller, A. and Kunze-Mühl, E. Zyclopie mit D-trisomie. Wien. Klin. Wochenschr. 84: 272-275 (1972).

39. Gorlin, R. J., Yunis, J. and Anderson, V. E. Short arm deletion of chromosome 18 in cebocephaly. Am. J. Dis. Child. 115: 473-476 (1968).

40. Gouw, W. L., ten Kate, L. P., Anders, G. J. P. A. and Okken, A. A case of 18q- in a family with a translocation t(6p+:18q-) identified by the Giemsa-banding technique. Humangenetik 19: 123-126 (1973).

41. Grouchy, J. de. The 18p-, 18q- and 18r syndromes. Birth Defects: Orig. Art. Ser. 5(pt. 5): 74-87 (1969).

42. Grouchy, J. de, Bonnette, J. and Salmon, C. Délétion du bras court du chromosome 18. Ann. Genet. (Paris) 9: 19-26 (1966).

43. Grouchy, J. de, Herrault, A. and Cohen-Solal, J. Une observation de chromosome 18 en anneau (18r). Ann. Genet. (Paris) 11: 33-38 (1968).

44. Grouchy, J. de, Lamy, M., Thieffry, S., Arthuis, M. and Salmon, C. Dymorphie complexe avec oligophrénie: Délétion des bras courts d'un chromosome 17-18. C. R. Acad. Sci. (Paris) 256: 1028-1029 (1963).

45. Grouchy, J. de, Lévêque, B., Debauchez, C., Salmon, Ch., Lamy, M. and Marie, J. Chromosome 17-18 en anneau et malformations congénitales chez une fille. Ann. Genet. (Paris) 7: 17-23 (1964a).

46. Grouchy, J. de, Royer, P., Salmon, C. and Lamy, M. Délétion partielle des bras longs du chromosome 18. Pathol. Biol. (Paris) 12: 579-582 (1964b).

47. Haenthens, P. Chromosome 18 en anneau. Arch. Fr. Pediatr. 24: 964 (1967).

48. Halbrecht, I., Kletzky, O., Komlos, L., Lotker, M. and Gersht, N. Trisomy-D in cyclops. Obstet. Gynecol. 37: 391-393 (1971).

49. Harlan, W. L., Cotton, J. E., Pettid, F., Fitzmaurice, F., Kaplan, A. R. and Lynch, H. T. Clinical manifestations in a child affected with mosaicism involving partial deletion of the short arm of an E-18 chromosome. In Proceedings of the Fourth International Congress of Human Genetics, J. de Grouchy, F. J. G. Ebling, I. Henderson, and J. François (Eds.), Amsterdam, Excerpta Medica (International Congress Series No. 233), 1971, p. 86.

50. Hoehn, H., Sander, C. and Sander, L. Z. Aneusomie de recombinaison: Rearrangement between paternal chromosomes 4 and 18 yielding offspring with features of the 18q- syndrome. Ann. Genet. (Paris) 14: 187-192 (1971).

51. Hooft, C., Haentjens, P., Orye, E., Kluyskens, F. and D'Hont, G. Chromosome 18 en anneau. Acta Paediatr. Belg. 22: 69-88 (1968).

52. Insley, J. Syndrome associated with a deficiency of part of the long arm of chromosome 18. Arch. Dis. Child. 42: 140-146 (1967).

53. Jacobsen, P. and Mikkelsen, M. Chromosome 18 abnormalities in a family with a translocation t(18p-:21p+). J. Ment. Defic. Res. 12: 144-161 (1968a).

54. Jacobsen, P. and Mikkelsen, M. The 18p- syndrome. Report of two cases. Ann. Genet. (Paris) 11: 211-216 (1968b).

55. Jacobsen, P., Mikkelsen, M., Niebuhr, E. and Grouchy, J. de. A case of 18q- in a t(18q-:6p+) family. Ann. Genet. (Paris) 14: 41-48 (1971).

56. Jeune, M., Lamit, J., Michel, M., Fournier, P., Laurent, C. and Nivelon, A. Un cas de chromosome 18 en anneau. Pediatrie 22: 845-846 (1967).

57. Kunze, J., Stephan, E. and Tolksdorf, M. Ring-Chromosom 18, ein 18p-/18q- Deletionssyndrom. Humangenetik 15: 289-318 (1972).

58. Lafourcade, J. and Lejeune, J. La déficience du bras long d'un chromosome 18 (18q-). Union Med. Can. 97: 936-940 (1968).

59. Laurent, C., Michel, M., Philippe, N. and Pinçon, J. A. Délétion du bras court du chromosome 18 et mosaique paternelle. Ann. Genet. (Paris) 13: 56-60 (1970).

60. Law, E. M. and Masterson, J. G. Familial 18q- syndrome. Ann. Genet. (Paris) 12: 215-222 (1969).

61. Law, E. M. and Masterson, J. G. Partial deletion of chromosome 18. Lancet 2: 1137 (1966).

62. Leisti, J., Leisti, S., Perheentupa, J., Savilahti, E. and Aula, P. Absence of IgA and growth hormone deficiency associated with short arm deletion of chromosome 18. Arch. Dis. Child. 48: 320-322 (1973).

63. Lejeune, J., Berger, R., Lafourcade, J. and Réthoré, M. -O. La délétion partielle du bras long du chromosome 18. Individualisation d'un nouvel état morbide. Ann. Genet. (Paris) 9: 32-38 (1966a).

64. Lejeune, J., Berger, R., Réthoré, M. -O., Paoline, P., Boisse, J. and Mozziconacci, P. Sur un cas de délétion partielle du bras court du chromosome 18, résultant d'une translocation familiale, 18c~17. Ann. Genet. (Paris) 9: 27-31 (1966b).

65. Lejeune, J., Berger, R., Réthoré, M. O. and Vialatte, J. A Case of the 18q- syndrome. J. Genet. Hum. 17: 423-428 (1969).

66. Lejeune, J., Réthoré, M. -O., Berger, R., Baheux, G. and Chabrune, J. Sur un cas de chromosome 18 en anneau. Ann. Genet. (Paris) 9: 173-175 (1966c).

67. Levenson, J. E., Crandall, B. F. and Sparkes, R. S. Partial deletion syndromes of chromosome 18. Case report and review of the literature. Ann. Ophthalmol. (Chic.), 3: 756-760 (1971).

68. Lurie, I. W. and Lazjuk, G. I. Partial monosomies 18. Humangenetik 15: 203-222 (1972).

69. McDermott, A., Insley, J., Barton, M. E., Rowe, P., Edwards, J. H. and Cameron, A. H. Arhinencephaly associated with a deficiency involving chromosome 18. J. Med. Genet. 5: 60-67 (1968).

70. Malpuech, G., Raynaud, E. J., Beun, J., Godeneche, P. and Grouchy, J. de. Délétion du bras court du 18 par translocation t(G-:18p+). Une étude en fluorescence par la moutarde de quinacrine. Ann. Genet. (Paris) 14: 213-218 (1971a).

71. Malpuech, G., Raynaud, E. -J., Gaulme, J. and Godeneche, P. Délétion complete du bras court du chromosome 18 et translocation G~18 avec dyschromie et hypothyroidie. Arch. Fr. Pediatr. 28: 837-846 (1971b).

72. Michaels, D. L., Go., S., Humbert, J. R., Dubois, R. S., Stewart, J. M. and Ellis, E. F. Intestinal nodular lymphoid hyperplasia, hypogammaglobulinemia, and hematologic abnormalities in a child with a ring 18 chromosome. J. Pediatr. 79: 81-88 (1971).

73. Migeon, B. R. Short arm deletions in group E and chromosomal "deletion" syndromes. J. Pediatr. 69: 432-438 (1966).

74. Mikelsaar, A. -V. N. and Talvik, T. A. Partial deletion of the long arm of chromosome 18. Humangenetik 7: 359-360 (1969).

75. Mikelsaar, A. -V. N. and Talvik, T. A. Syndrome of partial deletion of the long arm of chromosome 18 (18q-). Genetika 6(3): 162-170 (1970).

76. Mikelsaar, A. -V., Talvik, T. A. and Sitska, M. E. A ring-shaped chromosome (presumably No 18) and its bearing on the multiple congenital anomalies in man. Genetika 3(5): 63-66 (1967).

77. Moore, M. K. and Engel, E. Clinical, cytogenetic, and autoradiographic studies in 10 cases with rare chromosome disorders. III. Cases 6, 7 and 8. Ann. Genet. (Paris) 13: 207-212 (1970a).

78. Moore, M. K. and Engel, E. Clinical, cytogenetic and autoradiographic studies in 10 cases with rare chromosome disorders. IV. Cases 9 and 10. Ann. Genet. (Paris) 13: 269-274 (1970b).

79. Morič-Petrovič, S., Laća, Ž. and Kaličanin, P. Un cas de délétion du bras court d'un chromosome dans le groupe E (le syndrome de De Grouchy). In Proceedings of the Fourth International Congress of Human Genetics, J. de Grouchy, F. J. G. Ebling, I. Henderson, and J. François (Eds.), Amsterdam, Excerpta Medica (International Congress Series No. 233), 1971, p. 127.

80. Murken, J. -D., Salzer, G. and Kunze, D. Ring Chromosom Nr 18 und fehlendes IgA bei einen 6 jahrigen madchen (46, XX, 18r). Z. Kinderheilkd. 109: 1-10 (1970).

81. Nance, W. E., Higdon, S. H., Chown, B. and Engel, E. Partial E-18 long-arm deletion. Lancet 1: 303 (1968).

82. Neu, R. L., Watanabe, N., Gardner, L. I. and Galvis, A. G. A single nasal orifice and severe intrauterine growth retardation in a case of 46, XX, 18r. Ann. Genet. (Paris) 14: 139-142 (1971).

83. Nitowski, H. M., Sindhvananda, N., Konigsberg, U. R. and Weinberg, T. Partial 18 monosomy in the cyclops malformation. Pediatrics 37: 260-269 (1966).

84. Palmer, C. G., Fareed, N. and Merritt, A. D. Ring chromosome 18 in a patient with multiple anomalies. J. Med. Genet. 4: 117-123 (1967).

85. Pergament, E., Pietra, G. C., Kadotani, T., Sato, H. and Berlow, S. A ring chromosome No 16 in an infant with primary hypoparathyroidism. J. Pediatr. 76: 745-751 (1970).

86. Petit, P. and Poncelet, R. Un nouveau cas de chromosome 18 en anneau (18r) (1). Ann. Genet. (Paris) 10: 134-137 (1967).

87. Pfeiffer, R. A. Deletion den kurzen Arme des Chromosoms Nr 18. Humangenetik 2: 178-185 (1966).

88. Pfitzer, P. Extra minute chromosome in cyclops. Lancet 2: 102-103 (1967).

89. Picciano, D. J., Berlin, C. M., Davenport, S. L. H. and Jacobson, C. B. Human ring chromosomes: A report of five cases. Ann. Genet. (Paris) 15: 241-247 (1972).

90. Reinwein, H., Ritter, H. and Wolf, U. Deletion of short arm of a chromosome 18 (46,XX,18p-). Humangenetik 5: 72-73 (1967).

91. Richards, B. W., Rundle, A. T., Zaremba, J. and Stewart, A. Ring chromosome 18 in a mentally retarded boy. J. Ment. Defic. Res. 14: 174-186 (1970).

92. Ruvelcaba, R. H. A. and Thuline, H. C. IgA absence associated with short arm deletion of chromosome No 18. J. Pediatr. 74: 964-965 (1969).

93. Sanroman, C. and Réthoré, M. -O. Sindrome (18q-). Una nueva observacion en mosaico (46,XX/46,XX(18q-)). Rev. Clin. Esp. 114: 61-66 (1969).

94. Saraux, H. Types et contres-types en pathologie chromosomique. Bull. Mem. Soc. Fr. Ophtalmol. 85: 8-16 (1972).

95. Schinzel, A., Hayashi, K. and Schmid, W. Structural aberrations of chromosome 18. II. The 18q- syndrome. Report of 3 cases. Humangenetik 26: 123-132 (1975).

96. Schwanitz, G., Rott, H. D., Koch, G. and Gumminger, G. Chromosomal bedingte Missbildungssyndrome. Kasuistischer Beitrag zur Defizienz der kurzen Arme eines Chromosoms Nr 18 (46,XY,18p-). Med. Welt 20(3): 1708-1710 (1969).

97. Stewart, J. M., Go, S., Ellis, E. and Robinson, A. Absent IgA and deletions of chromosome 18. J. Med. Genet. 7: 11-19 (1970).

98. Šubrt, I. and Beránková, J. A case of the 18p- syndrome. Humangenetik 16: 359-360 (1972).

99. Summitt, R. L. Deletion of the short arm of chromosome 18. Cytogenetics 3: 201-206 (1964).

100. Taysi, K. and Tinaztepe, K. Trisomy D and the cyclops malformation. Am. J. Dis. Child. 124: 710-713 (1972).

101. Thieffry, S., Arthuis, M., Grouchy, J. de, Lamy, M. and Salmon, C. Délétion des bras courts d'un chromosome 17-18: Dysmorphies complexes avec oligophrénie. Arch. Fr. Pediatr. 20: 740-745 (1963).

102. Toews, H. A. and Jones, H. W. Cyclopia in association with D trisomy and gonadal agenesis. Am. J. Obstet. Gynecol. 102: 53-56 (1968).

103. Uchida, I. A., McRae, K. N., Wang, H. C. and Ray, M. Familial short arm deficiency of chromosome 18 concomitant with arhinencephaly and alopecia congenita. Am. J. Hum. Genet. 17: 410-419 (1965).

104. Vaillaud, J. C., Martin, J. and Ayraud, N. Un nouveau cas de délétion partielle du bras court du chromosome 18. Ann. Genet. (Paris) 13: 120-122 (1970).

105. Valdmanis, A., Pearson, G., Siegel, A. E., Hoeksema, R. H. and Mann, J. D. A pedigree of 4/18 translocation chromosomes with type and countertype partial trisomy and partial monosomy for chromosome 18. Ann. Genet. (Paris) 10: 159-166 (1967).

106. Wald, S., Engel, E., Nance, W. E., Davies, J., Puyau, F. A. and Sinclair-Smith, B. C. E ring chromosome with persistent left superior vena cava and hypertrophic subaortic stenosis. J. Med. Genet. 6: 328-333 (1969).

107. Wang, H. C., Melnyk, J., McDonald, L. T., Uchida, I. A., Carr, D. H. and Goldberg, B. Ring chromosomes in human beings. Nature (Lond.) 195: 733-734 (1962).

108. Wertelecki, W. and Gerald, P. S. Clinical and chromosomal studies of the 18q- syndrome. J. Pediatr. 78: 45-52 (1971).

109. Wertelecki, W., Schindler, A. M. and Gerald, P. S. Partial deletion of chromosome 18. Lancet 2: 641 (1966).

110. Wolf, U., Reinwein, H., Gorman, L. Z. and Künzer, W. Deletion on long arm of a chromosome 18 (46,XX,18q-). Humangenetik 5: 70-71 (1967).

111. Yanoff, M., Rorke, L. B. and Niederer, B. S. Ocular and cerebral abnormalities in chromosome 18 deletion defect. Am. J. Ophthalmol. 70: 391-402 (1970).

CHAPTER 7

DEFICIENCIES OF GROUP F CHROMOSOMES

I. OCULAR ABNORMALITIES

There are very few references in the literature to abnormalities of group F chromosomes, to those due either to trisomies or to deletions. This, possibly, may be due to the difficulties in detecting any abnormality in a small chromosome, and group F chromosomes are the smallest. In 1972, three papers were published describing patients with F deletions associated with ocular abnormalities, and it may be that the new techniques have made possible the detection of these chromosomal abnormalities.

Ahmed (1) described a mentally retarded man with short stature, microcephaly, flat occiput, slanting eyes with epicanthic folds, and an Fq- chromosome. The only ocular abnormality mentioned by Faed et al. (2) was epicanthus in an 11-year-old boy, who had a ring F with mosaicism 46,XY,20r/46,XY. The third reference is that of de Grouchy et al. (3), reporting a ring F chromosome in a boy with antimongoloid slant, convergent strabismus, and microphthalmos, with other systemic abnormalities.

It is impossible to discuss the significance of these few findings which are included only out of academic interest.

REFERENCES

1. Ahmed, M. Long arm deletion of 19/20 chromosomes. Lancet 1: 451 (1972).

2. Faed, M., Morton, H. G. and Robertson, J. Ring F chromosome mosaicism (46, XY, 20r/46, XY) in an epileptic child with apparent haematological disease. J. Med. Genet. 9: 470-472 (1972).

3. de Grouchy, J., Plachot, M., Sebaoun, M. and Bouchard, R. Chromosome F en anneau (46, XY, Fr) chez un garçon multimalformé. Ann. Genet. (Paris) 15: 121-126 (1972).

CHAPTER 8

DEFICIENCIES OF GROUP G CHROMOSOMES

I. OCULAR ABNORMALITIES

The ocular abnormalities associated with various types of deletions of a group G chromosome are summarized in Table 8-1. This is a very heterogeneous group, even when the complete phenotype is considered, and there are several interesting aspects from a cytogenetic point of view.

The most common autosomal chromosomal abnormality is trisomy 21 and it is the only trisomy compatible with life and development, although recently there have been a few reports of trisomy of C group chromosomes. It is interesting therefore that the only cases of autosomal monosomy in live subjects should be of a G monosomy, usually assumed to be monosomy 21. There are six such cases and all with ocular abnormalities (1, 2, 15, 16, 17, 41).

It will be seen from Table 8-1 that there are a number of cases of partial monosomy G with or without mosaicism. There may be several cell lines, but usually there is one cell line with a deleted G chromosome and another with a ring G chromosome. This is due to ring chromosomes being unstable structures, particularly at meiosis, and are liable to be lost at this stage of cell division. The first report of partial monosomy G is that of Lejeune et al. (26) who described a child with antimongoloid slant, blepharochalasis, and persistence of pupillary membrane, as well as other systemic abnormalities which led them to consider this case as an example of the countertype of mongolism. In 1966, Reisman et al. (34) coined the term "antimongolism" to describe this syndrome since the clinical features seem to be the antithesis of those found in mongolism. From the ophthalmic point of view, this may be illustrated by the prevalence of antimongoloid slant and the absence of Brushfield spots in these cases. On the other hand there are exceptions, since cataracts (8, 25, 34), Brushfield spots (23), and strabismus and nystagmus, abnormalities commonly seen in trisomy 21, have been reported by various investigators. The antimongolism syndrome is characterized by the facial appearance with downward slanting palpebral fissures, blepharochalasis, micrognathia, protruding nose or broad bridge to the nose, and large low-set ears. There are also severe cardiovascular, renal, and genito-urinary anomalies. It has been suggested by Warren and Rimoin (42) that there are two G deletion syndromes, GI and GII, with the GI syndrome corresponding to antimongolism. It was further suggested (28) that GI is due to a 21r chromosome and GII to a 22r. It is probable that several of the earlier cases with a G-/Gr karyotype correspond to a GI or GII syndrome, as they are partly G- due to the instability of the ring G chromosome.

The GII syndrome is characterized by mental retardation, ptosis, epicanthic folds, flat nasal bridge, and minor anomalies of the musculoskeletal system. GI is a more severe syndrome than GII, but they both have relatively minor ocular abnormalities. So far, ptosis has only been reported with GII.

It is possible to compare trisomy 21 with GI, the two having a completely different facial appearance, although there is mental retardation in both, cardiopathy in trisomy 21 and cardiovascular abnormalities in GI. The ocular abnormalities in each have been mentioned, cataracts seem to be a rare finding in GI but common in trisomy 21.

It is somewhat more difficult to compare trisomy 22 with GII. The evidence for the existence of trisomy 22 is not conclusive; most of the cases cited by Hamerton (18) were published before banding techniques. There is, however, one recent case of a probable trisomy 22 (14) identified by trypsin-Giemsa banding in a child with a coloboma of the right iris. A few of the children mentioned by Bass et al. (4) died in the first two years of life, so that trisomy 22, if it exists in live subjects, is a severe syndrome. Its corresponding deletion syndrome GII which is due to a 22r, causes only minor abnormalities, and on this basis one would expect trisomy 22 to be minor. There seems to be no references to a monosomy 22, with or without mosaicism. It may be that the case of cyclopia (10) was due to a monosomy 21, and that most of the other cases cited with a monosomy G were actually due to a monosomy 21. These are purely hypothetical considerations and will have to await further reports for confirmation.

TABLE 8-1

GROUP G CHROMOSOMES: DELETIONS AND RINGS

Author	Ocular Abnormalities	Chromosome Abnormality
Lejeune et al. (26) 1964	Antimongoloid slant, blepharochalasis, pupillary membrane	45, XY, G-/45, XY, Gr
Wolf et al. (46) 1964	Typically mongol eyes	Satellited large acrocentric with abnormally long short arms
Cohen (10) 1966	Cyclopia	45, XY, G-/46, XY
German and Bearn (13) 1966	Mongoloid slant	? Gr
Reisman et al. (34) 1966	Blepharoconjunctivitis, antimongoloid slant, cataracts	Partial monosomy 21
Thorburn and Johnson (41) 1966	Antimongoloid slant	Monosomy G
Al-Aish et al. (2) 1967	Hypertelorism, epicanthus, antimongoloid slant	Monosomy 21
Hall et al. (16) 1967	Antimongoloid slant, blepharochalasis, eccentric pupils	Monosomy G
Hecht et al. (19) 1967	Ptosis, epicanthus	Gr
Hoefnagel et al. (20) 1967	Epicanthus	Gr
Reisman et al. (33) 1967	Hypertelorism, epicanthus	Partial monosomy G
Bauchinger et al. (5) 1968	L. E. convergent strabismus, coloboma, phthisis bulbi	45, XY, G-/46, XY, Gcen(p-, q-)

TABLE 8-1 (continued)

Author	Ocular Abnormalities	Chromosome Abnormality
Schultz and Krmpotic (36) 1968	Case 1: large eyes, hypertelorism; case 2: epicanthus, hypertelorism	Monosomy G mosaic
Weleber et al. (45) 1968	Ptosis, epicanthus, violaceous lids	Gr
Al-Aish (1) 1969	Wide-set slanting eyes, epicanthus	Monosomy 21
Blank and Lorber (6) 1969	Slight epicanthus, mongoloid slant, no Brushfield spots	45, XX, G-/46, XX, Gr
Böhm and Fuhrmann (7) 1969	Epicanthus	45, XX, G-/46, XX
Challacombe and Taylor (8) 1969	Antimongoloid slant, cataract (LE), bilateral corneal opacities	45, XY, G-/46, XY, Gq- or Gr
Endo et al. (12) 1969	Bilateral cataract	45, XY, G-/46, XY, Gq-
Mikkelsaar (29) 1969	Slight epicanthus, hypertelorism	Gp-
Talvik and Mikkelsaar (40) 1969	Epicanthus	Gr
Zdansky et al. (47) 1969	Slight divergent strabismus, Brushfield spots	46, XY, Gr/45, XY, G-
Emberger et al. (11) 1970	Blepharochalasis, hypotelorism	45, XX, 21-/46, XX, 21pi
Say et al. (35) 1970	Antimongoloid slant, horizontal nystagmus	45, XY, G-/46, XY/46, XY, Gr
Singer and Scaife (38) 1970	Hypotelorism, mongoloid slant, normal fundi	Gr and inv(B)

TABLE 8-1 (continued)

Author	Ocular Abnormalities	Chromosome Abnormality
Warren and Rimoin (42) 1970	Case 1: epicanthus; case 2: antimongoloid slant	Gr (GII syndrome) Gr (GI syndrome)
Armendares et al. (3) 1971	Antimongoloid slant, hypotelorism	45, XY, G-/46, XY, Gr
Greenwood and Sommer (14) 1971	Antimongoloid slant, ptosis, R corneal opacity	45, XX, G-/46, XX in blood and 46, XX in skin
Kaijser (22) 1971	Antimongoloid slant	46, XY, G-?, F+
Kelch et al. (23) 1971	Case 1: antimongoloid slant, hypertelorism, blepharochalasis, Brushfield spots; case 2: epicanthus, hypertelorism, microphthalmos	46, XX, Gp-/45, XX, Gp- 46, XY, Gq-/45, XY, Gq-
Nevin et al. (31) 1971	Antimongoloid slant	46, XX, Gr
Weber et al. (44) 1971	Antimongoloid slant	45, X/45, XX, 21-
Chauvel et al. (9) 1972	Epicanthus, slight ptosis	45, XY, G-/46, XY, Gr GII syndrome
Gripenberg et al. (15) 1972	Slight antimongoloid slant	45, XX, 21-
Magenis et al. (28) 1972	Case 1: see (3); case 2: see (45)	46, XY, 21r (GI syndrome) 46, XX, 22r (GII syndrome)
Picciano et al. (32) 1972	Small eyes, epicanthus, strabismus	Gr
Koivisto et al. (24) 1973	Slight hypertelorism, nystagmus	3 group G chromosomes, metacentric fragment
Lindenbaum et al. (27) 1973	Mild epicanthus	22r

TABLE 8-1 (continued)

Author	Ocular Abnormalities	Chromosome Abnormality
Shibata et al. (37) 1973	Antimongoloid slant, convergent strabismus	45, XX, 21-/46, XX, 21r
Stoll et al. (39) 1973	Antimongoloid slant, epicanthus, hypertelorism, long eyelashes	22r
Warren et al. (43) 1973	Case 1: antimongoloid slant; case 2 and 3: ptosis, epicanthus	21r
Halloran et al. (17) 1974	Antimongoloid slant	45, XX, 21-
Holbeck et al. (21) 1974	LE: microphthalmos with grey and cloudy cornea	Monosomy for centromeric and juxtacentromeric region of 21
Kučerová and Polívková (25) 1974	Hypertelorism, antimongoloid slant, bilateral cataracts	46, XX, r(21)
Mikkelsen and Vestermark (30) 1974	Antimongoloid slant, blepharochalasis, ophthalmoscopy normal	45, XX, 21-/46, XX, 21q-

REFERENCES

1. Al-Aish, M. S. Aneuploidy of the G(21-22) autosomes, clinical and cytological approach. Birth Defects: Orig. Art. Ser. 5(pt. 5): 59-63 (1969).

2. Al-Aish, M. S., Cruz, F. de la, Goldsmith, L. A., Volpe, J., Mella, G. and Robinson, J. C. Autosomal monosomy in man: Complete monosomy G(21-22) in a four-and-one-half-year-old mentally retarded girl. N. Engl. J. Med. 277: 777-784 (1967).

3. Armendares, S., Buentello, L. and Cantu-Garza, J. -M. Partial monosomy of a G group chromosome (45,XY,G-/46,XY,Gr): Report of a new case. Ann. Genet. (Paris) 14: 7-12 (1971).

4. Bass, H. N., Crandall, B. F. and Sparkes, R. S. Probable trisomy 22 identified by fluorescent and trypsin-Giemsa banding. Ann. Genet. (Paris) 16: 189-192 (1973).

5. Bauchinger, M., Schmid, E. and Röttinger, E. Ein Fall mit einem Mosaik partielle Monosomie G/Monosomie G in den Lymphocyten des peripheren Blutes. Humangenetik 6: 303-310 (1968).

6. Blank, C. E. and Lorber, J. A patient with 45,XX,G-/46,XX,Gr mosaicism. J. Med. Genet. 6: 220-223 (1969).

7. Böhm, R. and Fuhrmann, W. Lebensfahigkeit bei Monosomie G. Monatsschr. Kinderheilkd. 117(4): 184-187 (1969).

8. Challacombe, D. N. and Taylor, A. Monosomy for a G autosome. Arch. Dis. Child. 44: 113-119 (1969).

9. Chauvel, P. J., Schindeler, J. D. and Warren, R. J. G deletion syndrome II. Humangenetik 14: 164-166 (1972).

10. Cohen, M. M. Chromosomal mosaicism associated with a case of cyclopia. J. Pediatr. 69: 793-798 (1966).

11. Emberger, J. -M., Rey, J., Rieu, D., Dossa, D., Bonnet, H. and Jean, R. Monosomie 21 avec mosaique 45, XX, 21-/46, XX, 21pi. Arch. Fr. Pediatr. 27: 1069-1079 (1970).

12. Endo, A., Yamamoto, M., Watanabe, G. -I., Suzuki, Y. and Sakai, K. "Antimongolism" syndrome. Br. Med. J. 4: 148-149 (1969).

13. German, J. L. and Bearn, A. G. Antimongolism: Personal communication cited by L. S. Penrose. Lancet 1: 497 (1966).

14. Greenwood, R. D. and Sommer, A. Monosomy G: Case report and review of the literature. J. Med. Genet. 8: 496-500 (1971).

15. Gripenberg, U., Elfving, J. and Gripenberg, L. A 45, XX, 21- child: Attempt at a cytological and clinical interpretation of the karyotype. J. Med. Genet. 9: 110-114 (1972).

16. Hall, B., Fredga, K. and Svenningsen, N. A case of monosomy G? Hereditas 57: 356-364 (1967).

17. Halloran, K. H., Breg, W. R. and Mahoney, M. J. 21 monosomy in a retarded female infant. J. Med. Genet. 11: 386-389 (1974).

18. Hamerton, J. L. Human Cytogenetics, Vol. II. New York, Academic Press, 1971, pp. 262-267.

19. Hecht, F., Weleber, R. G. and Giblett, E. R. Chromosome anomalies. Lancet 1: 848 (1967).

20. Hoefnagel, D., Schroeder, T. M., Benirschke, K. and Allen, F. H. A child with a group-G ring chromosome. Humangenetik 4: 52 (1967).

21. Holbek, S., Friedrich, U., Brostrøm, K. and Petersen, G. B. Monosomy for the centromeric and juxtacentromeric region of chromosome 21. Humangenetik 24: 191-195 (1974).

22. Kaijser, K. Heart malformations in two brothers with identical chromosome aberrations (46, XY, G-? F+). Clin. Genet. 2: 255-260 (1971).

23. Kelch, R. P., Franklin, M. and Schmickel, R. D. Group G deletion syndromes. J. Med. Genet. 8: 341-345 (1971).

24. Koivisto, M., Schröeder, J. and Chapelle, A. de la. Probable monosomy-21 and partial trisomy. Am. J. Dis. Child. 125: 426-428 (1973).

25. Kučerová, M. and Polívková, Z. A case of a girl with a 21 ring chromosome. Hum. Hered. 24: 100-104 (1974).

26. Lejeune, J., Berger, R., Réthoré, M., Archambault, L., Jérôme, H., Thieffry, S., Arcadi, J., Broyer, M., Lafourcade, J., Cruveiller, J. and Turpin, R. Monosomie partielle pour un petit acrocentrique. C. R. Acad. Sci. (Paris) 259: 4187-4190 (1964).

27. Lindenbaum, R. H., Bobrow, M. and Barber, L. Monozygotic twins with ring chromosome 22. J. Med. Genet. 10: 85-89 (1973).

28. Magenis, R. E., Armendares, S., Hecht, F., Weleber, R. G. and Overton, K. Identification by fluorescence of two G rings (46, XY, 21r) G deletion syndrome I and (46, XX, 22r) G deletion syndrome II. Ann. Genet. (Paris) 15: 265-266 (1972).

29. Mikelsaar, A. -V. N. Deletion of the short arm of one of the small acrocentric chromosomes of the G group. Genetika 5(4): 122-126 (1969).

30. Mikkelsen, M. and Vestermark, S. Karyotype 45, XX, -21/46, XX, 21q- in an infant with symptoms of G- deletion syndrome I. J. Med. Genet. 11: 389-392 (1974).

31. Nevin, N. C., MacLaverty, B. and Campbell, W. A. B. A child with a ring G chromosome (46, XX, Gr). J. Med. Genet. 8: 231-234 (1971).

32. Picciano, D. J., Berlin, C. M., Davenport, S. L. H. and Jacobson, C. B. Human ring chromosomes: A report of five cases. Ann. Genet. (Paris) 15: 241-247 (1972).

33. Reisman, L. E., Darnell, A., Murphy, J. W., Hall, B. and Kasahara, S. A child with partial deletion of a G-group autosome. Am. J. Dis. Child. 114: 336-339 (1967).

34. Reisman, L. E., Kasahara, S., Chung, C. Y., Darnell, A. and Hall, B. Anti-mongolism studies in an infant with a partial monosomy of the 21-chromosome. Lancet 1: 394-397 (1966).

35. Say, B., Tunçbilek, E., Yamak, B. and Balci, S. An unusual chromosomal aberration in a case of Chediak-Higashi syndrome. J. Med. Genet. 7: 417-421 (1970).

36. Schulz, J. and Krmpotic, E. Monosomy G mosaicism in two unrelated children. J. Ment. Defic. Res. 12: 255-268 (1968).

37. Shibata, K., Waldenmaier, C. and Hirsch, W. A child with a 21-ring chromosome, 45, XX, 21-/46, XX, 21r, investigated with the banding technique. Humangenetik 18: 315-319 (1973).

38. Singer, H. and Scaife, N. S. Simultaneous occurrence of ring "G" chromosome and group "B" pericentric inversion in the same individual: Case report and review of the literature. Pediatrics 46: 74-83 (1970).

39. Stoll, C., Rohmer, A. and Sauvage, P. Chromosome 22 en anneau r(22): Identification par dénaturation thermique ménagée. Ann. Genet. (Paris) 16: 193-198 (1973).

40. Talvik, T. A. and Mikelsaar, A. -V. N. A new case of ring chromosome of the group 21-22 (G_R?). Genetika 5(12): 129-133 (1969).

41. Thorburn, M. J. and Johnson, B. E. Apparent monosomy of a G autosome in a Jamaican infant. J. Med. Genet. 3: 290-292 (1966).

42. Warren, R. J. and Rimoin, D. L. The G-deletion syndrome. J. Pediatr. 77: 658-663 (1970).

43. Warren, R. J., Rimoin, D. L. and Summitt, R. L. Identification by fluorescent microscopy of the abnormal chromosomes associated with the G-deletion syndromes. Am. J. Hum. Genet. 25: 77-81 (1973).

44. Weber, F. M., Sparkes, R. S. and Muller, H. Double monosomy mosaicism (45X/45, XX, 21-) in a retarded child with multiple congenital malformations. Cytogenetics 10: 404-412 (1971).

45. Weleber, R. G., Hecht, F. and Giblett, E. R. Ring G chromosome: A new G-deletion syndrome. Am. J. Dis. Child. 115: 489 (1968).

46. Wolf, U., Baitsch, H., Künzer, W. and Reinwein, H. Familiares Auftreten eines anomalen D-Chromosoms. Cytogenetics 3: 112-123 (1964).

47. Zdansky, R., Bühler, E. M., Vest, M., Bühler, U. K. and Stalder, G. R. Familiares Mosaik mit G Ring. Humangenetik 7: 275-286 (1969).

CHAPTER 9

MONOSOMY X

I. TURNER SYNDROME AND TURNER MOSAICS

Turner (58) first described a syndrome with clinical features of infantilism, congenital webbed neck, and cubitus valgus. The essential defect in this syndrome is the presence of "streak gonads" which are fibrous tissue resembling ovarian stroma instead of normal ovaries. Clinically there may be several forms of this syndrome depending on whether ovarian dysgenesis is or is not associated with webbing of the neck and other anomalies. In the context of this review, the Turner syndrome will be defined as the association of ovarian dysgenesis with an XO karyotype.

A. Incidence

The 45, XO karyotype is the most common single chromosomal abnormality found in spontaneous abortions, occurring in about 5% of all spontaneous abortions or in 20% of all chromosomally abnormal fetuses (28). The incidence of 45, X and 45, X/46, XX is probably about 0.4 per 1,000 female births.

B. Phenotype

As has been stated, ovarian dysgenesis may be present with or without webbing of the neck. Other characteristic signs of the syndrome include: broad shield-shaped chest with widely spaced nipples; infantile development or absence of secondary sexual characteristics; short stature; a variety of skeletal anomalies; and abnormal structure of the hands.

There is also an increased frequency of cardiac and renal anomalies, and of metabolic disorders such as diabetes mellitus. The effect of ovarian dysgenesis at puberty is to produce absence of sexual development and primary amenorrhoea; it is extremely rare to find XO women who are menstruating regularly and who are fertile.

C. Cytogenetics

There are two categories of ovarian dysgenesis: (a) those which are chromatin-negative, and (b) those which are chromatin-positive. Hamerton (28) has summarized the findings of six major studies of patients with ovarian dysgenesis in which the chromosomes of 288 girls were studied in detail. Among the 180 girls who were chromatin-negative, approximately 90% had a 45, X chromosome complement and 10% were mixoploids with a 45, X cell line and an additional cell line which may be XX, XYY or XYdic. The other group consisted of 108 girls who were chromatin-positive, and in this group there were none with a 45, X complement, but instead, 80% were mixoploid, and under 20% showed a structurally abnormal X chromosome.

D. Ocular Abnormalities

From an ophthalmological point of view there is little to distinguish a typical Turner syndrome from a Turner mosaic, or a chromatin-positive from a chromatin-negative Turner syndrome. The ocular abnormalities described in the literature are summarized in Tables 9-1 and 9-2, and the most common ones found in both syndromes are compared in Table 9-3. It will be seen that the most frequent sign in the Turner syndrome is epicanthus; in this review it was found to occur in 16.2% of cases with Turner mosaic. While epicanthus is commonly found in association with other chromosomal abnormalities, and occurs frequently in normal children, it persists in only 2 to 3% of adults. The majority of patients with the Turner syndrome are ascertained in puberty or in adulthood, so that an incidence of 16.2% with epicanthus, a characteristic sign in the syndrome, is high. Hypertelorism also has a higher incidence in the Turner syndromes than in the normal adult population, and this abnormally high incidence is also true of ptosis, strabismus, and blue scleras. Cataracts have been reported in 16 cases of the Turner syndrome and in 8 cases of Turner mosaics, but the type of cataract is not usually specified. According to Lessell and Forbes (36), if diabetes is unduly frequent in the Turner syndrome, then cataracts can be expected more frequently and with earlier onset than in the general population.

It is generally accepted that the incidence of red-green colour-blindness in the Turner syndrome is the same as that in the normal male population. This is what one would expect, for the genes responsible for colour vision are located on the X chromosome. In this review, however, the incidence of defective colour vision is 3.1%, and this lower than expected incidence is probably due to all patients not having their colour vision tested. The incidence of 7.9% in Turner mosaics reflects a bias of selection. If a patient with the Turner syndrome has defective colour vision, the origin of her X chromosome can sometimes be ascertained, for if her father has normal colour vision, the patient's X chromosome must then be maternal in origin. If, however, her father has defective colour vision, the origin of her X chromosome will be paternal so long as her mother is not a heterozygote.

Other ocular abnormalities mentioned in the literature are nystagmus, corneal dystrophy, eccentric pupil, glaucoma, and retinal abnormalities. These are few in number and not specific. Certain cases are quite atypical of the Turner syndrome ophthalmologically. In the patient reported by Decourt et al. (14), the Turner syndrome was associated with albinism, depigmented irides and chorioretinitis with optic atrophy. It may be assumed that the typical ocular abnormalities in the Turner syndrome and Turner mosaics are epicanthus, strabismus, ptosis, cataracts, and defective colour vision. The ocular abnormalities associated with structural abnormalities of the X chromosomes are listed in Table 9-4, and they are very similar to those found in the Turner syndrome.

TABLE 9-1

Turner Syndrome

References	Ptosis	Epicanthus	Hypertelorism	Antimongoloid Slant	Strabismus	Nystagmus	Refractive Error	Abnormal Colour Vision	Blue Sclera	Corneal Dystrophy	Glaucoma	Cataract	Retinal Abnormality	Remarks
Ford et al. (24) 1959								Norm						
Laurent et al. (34) 1961											(+)			
Vaharu et al. (59) 1961				+	R Con									
de la Chapelle (9) 1962								1/13						Type of colour-blindness not stated
Becker et al. (2) 1963		+												Karyotype XX/Xx

Decourt et al. (14) 1963					Alt	+	+							Horizontal congenital nystagmus; depigmented iris; chorioretinitis with optic atrophy
Court-Brown et al. (13) 1964		+	+		Con			D						38 cases: Hypertelorism and epicanthus, 4; epicanthus, 4; convergent strabismus, 1; deuteranopia, 1
Ferguson-Smith et al. (19) 1964		4/30					1/30							Myopia 1/30; exophthalmos 1/30
Quinodoz et al. (51) 1964		1/6	1/6				1/6							Myopia and astigmatism 1/6; myopic conus and bilateral anomaly of retinal pigmentation 1/6
Engel and Forbes (17) 1965					7/16					2/16		3/16		
Hart et al. (29) 1965		+			Con									
Lessell and Forbes (36) 1966	6/37	2/37			7/37			1/37			1/37	3/37		Corneal nebulas 4/37
Benson et al. (13) 1967		+												

TABLE 9-1 (continued)

References	Ptosis	Epicanthus	Hypertelorism	Antimongoloid Slant	Strabismus	Nystagmus	Refractive Error	Abnormal Colour Vision	Blue Sclera	Corneal Dystrophy	Glaucoma	Cataract	Retinal Abnormality	Remarks
Calmettes et al. (6) 1968							M			+				Slight divergent strabismus; visual acuity 3/10 bilaterally
Fischer and Haslund (22) 1968					Con		M							Iris heterochromia; visual acuity 6/36 bilaterally
Hooft et al. (30) 1968		+		+										Pericentric inversion, 45,X,2(p+q−)
Thomas et al. (57) 1969		6/30			3/30	1/30		+						Small oval corneas 18/30; eccentric pupil 15/30; defective colour vision 8% (sic); blue scleras 3% (sic); retinitis pigmentosa 1/30

Bove (4) 1970		+												
Casteels-Vandaele et al. (8) 1970														Slight ptosis and epicanthus
Gardner (25) 1970 Case 1					+									
Laurent et al. (33) 1970		+												Familial t(Cq-:Gp+)
Nora et al. (46) 1970		5/16	7/16											
Say et al. (54) 1971														Mild exophthalmos
Wesson (60) 1971				2/11	9/11			3/9						
Peyresblanques and Chaban (48) 1972											+		+	Diabetic retinopathy
Sarto (53) 1974		2/9												

TABLE 9-2

Turner Syndrome XX/XO Mosaics

References	Ptosis	Epicanthus	Hypertelorism	Antimongoloid Slant	Strabismus	Nystagmus	Refractive Error	Abnormal Colour Vision	Blue Sclera	Corneal Dystrophy Glaucoma	Cataract	Retinal Abnormality	Remarks
Carr et al. (7) 1962								+					Minimal red-green deficiency XO/XX/XXX
Ferrier et al. (21) 1962 Case 2					+								XO/Xx
Briggs et al. (5) 1963					+								XO/XX/XXX
Lindsten et al. (38) 1963b								+					XO/XXi
Mikkelsen et al. (44) 1963		+	+										XO/XX

London et al. (39) 1964								P			+		Bilateral anterior polar lens opacities XX/XO
Court-Brown et al. (13) 1964	+	+	+			+		P					XO/XX (11 cases): colour vision normal in all; XO/XXX (3 cases): colour vision normal in all; XO/XY (5 cases): protanopia (1); hypertelorism (1)
Engel and Forbes (17) 1965	3/13				3/13		2/13		2/3		4/13		XO/XX
Cordier et al. (12) 1966					Div						+	+	Retinitis pigmentosa XO/XX
Jagiello et al. (31) 1966			2/2										XO/XY case 12; XX/XO/Xr case 18
Leao et al. (35) 1966		1/9	1/9		1/9								Exopthalmos 1/9 XO/XX
Pierson et al. (49) 1966											2/7		Eccentric pupil 4/7; XO/XXqi (2); XO/Xf (3); XXqi (2)
Cordier et al. (11) 1967								D					Deuteranopia in both cases: XO/X_F; XY/XO
Neuhäuser and Back (45) 1968 Case 2					+								Microcornea; XO/XXcen
Case 3													Proptosis; XO/XXcen

TABLE 9-2 (continued)

References	Ptosis	Epicanthus	Hypertelorism	Antimongoloid Slant	Strabismus	Nystagmus	Refractive Error	Abnormal Colour Vision	Blue Sclera	Corneal Dystrophy	Glaucoma	Cataract	Retinal Abnormality	Remarks
Aubert et al. (1) 1969					+									XO/XX
Ferguson-Smith et al. (20) 1969														Moderate hypertensive retinopathy; XO/XYqi
Predescu et al. (50) 1969		+												XO/XX

Giraud et al. (26) 1970														Bilateral macular lesion; pigmentary "powdering" of the retina: XO/XY
Nora et al. (46) 1970	3/4	2/4												XO/XX
Jancar (32) 1970								P					+	Retinitis pigmentosa: XO/XX
Mikelsaar et al. (43) 1971b														"Ocular infections": XO/XX
Eller (16) 1971		1/6	3/6											XO/XX

TABLE 9-3

Comparison of Ocular Findings in XO and XO/XX Syndromes

Ocular Abnormality	XO (222 Cases)	XO/XX (76 Cases)
Ptosis	11 (5.0%)	7 (9.2%)
Epicanthus	36 (16.2%)	7 (9.2%)
Hypertelorism	2 (0.9%)	8 (10.5%)
Antimongoloid slant	4 (1.8%)	0
Strabismus	32 (14.4%)	9 (11.8%)
Nystagmus	2 (0.9%)	1 (1.3%)
Refractive Error	5 (2.2%)	2 (2.6%)
Abnormal colour vision	7 (3.1%)	6 (7.9%)
Blue sclera	6+ (3%)	2 (2.6%)
Corneal dystrophy	3 (1.4%)	0
Glaucoma	3 (1.4%)	0
Cataract	16 (7.2%)	8 (10.5%)
Retinal abnormality	3 (1.4%)	4 (5.3%)

TABLE 9-4

Ocular Abnormalities Associated with Structural Abnormalities of the X Chromosome

Author	Ocular Abnormalities	X Chromosome Abnormality
Engel et al. (18) 1962	"Puffiness about the eyes"	XXqi
Lindsten et al. (37) 1963a	Five families studied, and in one family, one patient said to be colour blind and one deuteranope	XXqi
Fisher (23) 1965	Bilateral epicanthus, slow rotatory nystagmus, apparently blind	Mosaic XX/XO and XX/?Xr
Ree (52) 1965	Vision 6/12 bilaterally	XXqi
Paolini et al. (47) 1966	Partial red-green colour blindness	XO/XXr
Cohen et al. (10) 1967	One case with asymmetrical features; left half less developed than right; slight antimongoloid slant	XXr and centric fragment
Deminatti et al. (15) 1968	Hypertelorism	XXr
Melin and Samuelson (41) 1969	Colour amblyopia, green-red quotient between 2 and 1.3	XXq-
Luciani et al. (40) 1971	Slight hypertelorism, blue scleras	XXr
Mikelsaar et al. (42) 1971a	Hypertelorism, epicanthus, exophthalmos and speckled iris	47, XX, 21+/ 47, XXp-q-, 21+
Talvik et al. (56) 1971	Epicanthus	Assumed XXr/XO

TABLE 9-4 (continued)

Author	Ocular Abnormalities	X Chromosome Abnormality
Senzer et al. (55) 1973	Epicanthus 1/6, refractive error 4/6, strabismus 5/6	XXqi
Giraud et al. (27) 1974	Convergent strabismus (RE) in 17 year old	XXp-
Sarto (53) 1974	Epicanthus (2 cases)	45, X/46Xi(Xq) 46, Xi(Xq)

REFERENCES

1. Aubert, L., Arroyo, H., Vougier, M., Stahl, A., Carlon, N. and Franchimont, P. Nanisme essentiel et mosaique XO/XX. Presse Med. 77: 1051-1052 (1969).

2. Becker, K. L., Hayles, A. B. and Albert, A. Gonadal Dysgenesis due to mosaicism of an X chromosome fragment. Mayo Clin. Proc. 38: 422-426 (1963).

3. Benson, P. F., Taylor, A. I. and Gough, M. H. Chromosome anomalies in primary lymphoedema. Lancet 1: 461-462 (1967).

4. Bove, K. E. Gonadal dysgenesis in a newborn with XO karyotype. Am. J. Dis. Child. 120: 363-366 (1970).

5. Briggs, D. K., Stimson, C. W. and Vinograd, J. Leucocyte anomaly, mental retardation, and dwarfism in a family with abnormal chromosomes. J. Pediatr. 63: 21-28 (1963).

6. Calmettes, L., Deodati, F., Bec, P. and Labro, J. -B. Oxycéphalie et dystrophie cornéenne chez une malade présentant un syndrome de Turner. Bull. Soc. Ophtalmol. Fr. 68: 358-362 (1968).

7. Carr, D. H., Morishima, A., Barr, M. L., Grumbach, M. M., Luers, T. and Boschann, H. W. An XO/XX/XXX mosaicism in relationship to gonadal dysgenesis in females. J. Clin. Endocrinol. Metab. 22: 671-677 (1962).

8. Casteels-Vandaele, M., Proesmans, W., Berghe, H. van den and Verresen. Down's anomaly (21 trisomy) and Turner's syndrome (46, XXqi) in the same sibship. Helv. Paediatr. Acta 25: 412-420 (1970).

9. Chapelle, A. de la. Cytogenetical and clinical observations in female gonadal dysgenesis. Acta Endocrinol. (Kbh) 40(suppl. 65): 1-122 (1962).

10. Cohen, M. M., Sandberg, A. A., Takagi, N. and MacGillivray, M. H. Autoradiographic investigations of centric fragments and rings in patients with stigmata of gonadal dysgenesis. Cytogenetics 6: 254-267 (1967).

11. Cordier, J., Gilgenkrantz, S., Reny, A. and Raspiller, A. Syndrome de Turner et dyschromatopsie. Bull. Soc. Ophtalmol. Fr. 67: 1129-1131 (1967).

12. Cordier, J., Tridon, P. and Reny, A. Syndrome de Turner et rétinite pigmentaire. J. Genet. Hum. (Suppl. 15): 105-108 (1966).

13. Court-Brown, W. M., Harnden, D. G., Jacobs, P. A., Maclean, N. and Mantle, D. J. Abnormalities of the Sex Chromosome Complement in Man. MRC Special Report Series 305, London, HM Stationery Office, 1964.

14. Decourt, J., Michard, J. P. and Saltiel, H. Syndrome de Turner associé à un albinisme. Ann. Endocrinol. (Paris) 24: 866-869 (1963).

15. Deminatti, M., Maillard, E., Fossati, P. and Bulteel, M. -F. A propos d'un cas de chromosome X en anneau. Ann. Genet. (Paris) 11: 56-58 (1968).

16. Eller, E., Frankenburg, W., Puck, M. and Robinson, A. Prognosis in newborn infants with X-chromosomal abnormalities. Pediatrics 47: 681-688 (1971).

17. Engel, E. and Forbes, A. P. Cytogenetic and clinical findings in 48 patients with congenitally defective or absent ovaries. Medicine (Baltimore) 44: 135-164 (1965).

18. Engel, E., Northcutt, R. C. and Bunting, K. W. Diabetes and hypothyroidism with thyroid autoantibodies in a patient with a long arm X-isochromosome. J. Clin. Endocrinol. 29: 130-132 (1962).

19. Ferguson-Smith, M. A., Alexander, D. S., Bowen, P., Goodman, R. M., Kaufman, B. N., Jones, H. W., Jr. and Heller, R. H. Clinical and cytogenetic studies in female gonadal dysgenesis and their bearing on the cause of Turner's syndrome. Cytogenetics 3: 355-383 (1964).

20. Ferguson-Smith, M. A., Boyd, E., Ferguson-Smith, M. E., Pritchard, J. G., Yusuf, A. F. M. and Gray, B. Isochromosome for long arm of Y chromosome in patient with Turner's syndrome and sex chromosome mosaicism (45,X/46,XYqi). J. Med. Genet. 6: 422-425 (1969).

21. Ferrier, P., Gartler, S. M., Waxman, S. H. and Shepard, T. H. Abnormal sexual development associated with sex chromosome mosaicism. Pediatrics 29: 703-713 (1962).

22. Fischer, M. and Haslund, J. Severe mental retardation in Turner's syndrome and an additional mosaic with a centric chromosome fragment. Acta Genet. (Basel) 18: 487-495 (1968).

23. Fisher, G. W. Ring chromosome mosaicism in a severely subnormal child with multiple congenital malformations. J. Ment. Defic. Res. 9: 39-50 (1965).

24. Ford, C. E., Jones, K. W., Polani, P. E., DeAlmeida, J. C. and Briggs, J. H. A Sex-chromosome anomaly in a case of gonadal dysgenesis (Turner's syndrome). Lancet 1: 711-713 (1959).

25. Gardner, L. I. Pseudo-pseudohypoparathyroidism due to unequal crossing-over. Lancet 2: 879-880 (1970).

26. Giraud, F., Hartung, M., Brusquet, Y., Mattei, J. F., Sebahoun, A., Serment, H., Stahl, A. and Bernard, R. La fertilité des femmes 45,X/46,XX et 45,X/46,XX/47,XXX. Ann. Genet. (Paris) 13: 255-258 (1970).

27. Giraud, F., Hartung, M., Mattei, J. -F., Bachelet, Y. and Mattei, M. -G. Délétion partielle du bras court d'un chromosome X. Arch. Fr. Pediatr. 31: 717-724 (1974).

28. Hamerton, J. L. Human Cytogenetics, New York, Academic Press Inc. 1971, vol. II, pp. 82-89.

29. Hart, Z. H., Cohen, M. M., Dietze, M. R. and Reisman, L. E. A sex chromatin negative individual with chromosomes (XO) plus a persistent centric fragment. J. Pediatr. 66: 120-123 (1965).

30. Hooft, C., Coetsier, H. and Orye, E. Syndrome de Turner et inversion péricentrique probable du chromosome No 2 [45, X, 2 (p+q-)]. Ann. Genet. (Paris) 11: 181-183 (1968).

31. Jagiello, G. M., Kaminetsky, H. A., Ricks, Ph., Jr. and Ryan, R. J. Primary amenorrhea. A cytological and endocrinologic study of 18 cases. JAMA 198: 30-38 (1966).

32. Jancar, J. Retinitis pigmentosa with mental retardation, deafness and XX/XO sex chromosomes. J. Ment. Defic. Res. 14: 269-273 (1970).

33. Laurent, C., Bonnet, P., Farouz, S. and Longin, B. Association d'un cas de syndrome de Turner à une t(Cq-:Gp+) familiale. Ann. Genet. (Paris) 13: 61-66 (1970).

34. Laurent, C., Royer, J. and Noël, G. Syndrome de Turner et glaucome congénital. Bull. Soc. Ophtalmol. Fr. 61: 367-369 (1961).

35. Leao, J. C., Voorhess, M. L., Schlegel, R. J. and Gardner, L. I. XX/XO Mosaicism in nine preadolescent girls: Short stature as a presenting complaint. Pediatrics 38: 972-981 (1966).

36. Lessell, S. and Forbes, A. P. Eye signs in Turner's syndrome. Arch. Ophthalmol. 76: 211-213 (1966).

37. Lindsten, J., Fraccaro, M., Ikkos, D., Kaijser, K., Klinger, H. P. and Luft, R. Presumptive isochromosomes for long arms of X in man: Analysis of five families. Ann. Hum. Genet. 26: 383-405 (1963a).

38. Lindsten, J., Fraccaro, M., Polani, P. E., Hamerton, J. L., Sanger, R. and Race, R. R. Evidence that the Xg blood group genes are on the short arm of the X chromosome. Nature (Lond.) 197: 648-649 (1963b).

39. London, D. R., Kemp, N. H., Ellis, J. R. and Mittwoch, U. Turner's syndrome with secondary amenorrhea and sex chromosome mosaicism. Acta Endocrinol. (Kbh) 46: 341-351 (1964).

40. Luciani, J. M., Arroyo, H., Carlon, N., Aubert, L. and Stahl, A. Syndrome de Turner avec chromosome X en anneau. Sem. Hop. Paris 47: 2977-2986 (1971).

41. Melin, K. and Samuelson, G. Gonadal dysgenesis with lymphocytic thyroiditis and deletion of the long arm of the X chromosome. Acta Paediatr. Scand. 58: 625-631 (1969).

42. Mikelsaar, A. -V. N., Blumina, M. G., Kuznetsova, L. I., Mikelsaar, R. -V. A. and Lurie, I. V. A double chromosome aberration 47,XX,21+/47,XXp-q-,21+ in a girl with symptoms of Down's and Turner's syndromes. Genetika 7(5): 156-161 (1971a).

43. Mikelsaar, A. -V. N., Naydionova, M. M., Talvik, T. A., Mikelsaar, R. V. -A., Kaselaid, V. L., Zucker, E. I. and Grinberg, K. N. XX/XO-mosaicism in a newborn female infant with an atypical phenotype. Genetika 7(2): 154-160 (1971b).

44. Mikkelsen, M., Frøland, A. and Ellebjerg, J. XO/XX mosaicism in a pair of presumably monosygotic twins with different phenotype. Cytogenetics 2: 86-98 (1963).

45. Neuhäuser, G. and Back, F. Zytogenetische Varianten des Ullrich-Turner-Syndroms. Med. Klin. 63: 836-841 (1968).

46. Nora, J. J., Torres, F. G., Sinha, A. K. and McNamara, D. G. Characteristic cardiovascular anomalies of XO Turner syndrome, XX and XY phenotype and XO/XX Turner mosaic. Am. J. Cardiol. 25: 639-641 (1970).

47. Paolini, P., Berger, R., Réthoré, M. -O., Lafourcade, J. and Lejeune, J. Sur un cas de chromosome X en anneau. Ann. Genet. (Paris) 9: 78-79 (1966).

48. Peyresblanques, J. and Chaban, J. Syndrome de Turner et diabète. Bull. Soc. Ophtalmol. Fr., 72: 647-654 (1972).

49. Pierson, M., Gilgenkrantz, S., Olive, D., Dalaut, J. J. and Schaak, J. -C. Anomalies de structure d'un chromosome sexuel dans sept cas de dysgénésie gonadique de type Turner. Arch. Fr. Pediatr. 23: 1135-1153 (1966).

50. Predescu, V., Christodorescu, D., Tavtu, C., Ciovirnache, M. and Constantinescu, E. Repeated abortions in a woman with XO/XX mosaicism. Lancet 2: 217 (1969).

51. Quinodoz, J. M., Ferrier, P., Ferrier, S., Zahnd, G. and Prod'hom, A. Le syndrome de Turner: A propos de six observations. Helv. Med. Acta 31: 1-28 (1964).

52. Ree, M. J. Ovarian dysgenesis and presumed isochromosome of the long arm of X. J. Med. Genet. 2: 205-211 (1965).

53. Sarto, G. E. Cytogenetics of fifty patients with primary amenorrhea. Am. J. Obstet. Gynecol. 119: 14-23 (1974).

54. Say, B., Balci, S. and Tunçbilek, E. 45,XO Turner's syndrome, Wilm's tumor and imperforate anus. Humangenetik 12: 348-350 (1971).

55. Senzer, N., Aceto, T., Cohen, M. M., Ehrhardt, A. A., Abbassi, V. and Capraro, V. J. Isochromosome X. Am. J. Dis. Child. 126: 312-316 (1973).

56. Talvik, T. A., Mikelsaar, R. V. A. and Kask, V. A. A ring heterosome (46,XXr or 46XYr) in a 25-year-old girl. Genetika 7(7): 136-142 (1971).

57. Thomas, Ch., Cordier, J. and Reny, A. Les manifestations ophtalmologiques du syndrome de Turner. Arch. Ophtalmol. (Paris) 29: 565-574 (1969).

58. Turner, H. H. A syndrome of infantilism, congenital webbed neck and cubitus valgus. Endocrinology 23: 566-574 (1938).

59. Vaharu, T., Patton, R. C., Voorhess, M. L. and Gardner, L. I. Gonadal dysplasia and enlarged phallus in a girl with 45 chromosomes plus "fragment". Lancet 1: 1351 (1961).

60. Wesson, M. E. Turner's syndrome. Am. Orthopt. J. 21: 50-58 (1971).

CHAPTER 10

DUPLICATIONS

I. PARTIAL TRISOMIES

The ocular abnormalities associated with partial trisomies are listed in Table 10-1. In all these reports, there is the usual wide variety of minor abnormalities of the lids and palpebral fissures, strabismus and nystagmus, which are associated with most chromosomal abnormalities. Even though these characteristics are all nonspecific, they have to be considered as part of the overall clinical picture which in some cases, such as the 4p+ syndrome, is well defined and documented. For instance, nearly all the children in these reports showed some degree of mental retardation which in some cases was severe. Partial trisomy may cause relatively minor physical abnormalities so that it is ascertained through mental retardation in children who are over ten years of age, and yet in other cases the congenital abnormalities are so severe that the child dies in the first few months of life. Many of the children with partial trisomy have abnormal dermatoglyphs, but this particular aspect has been omitted from this review entirely as it is a highly specialized topic. It should be noted, however, that many of these children have abnormalities of the hands and fingers. Almost all the duplications in this section have arisen from the segregation of a balanced translocation or from a parental inversion. These are rare syndromes which were not recognized until the identification of the chromosome segment in excess was made possible with banding techniques. There are references in the literature to abnormal chromosomes, but these have been omitted unless the origin of the duplicated segment is known.

A. Group A

There are eight references in the literature to ocular abnormalities associated with partial trisomy of group A chromosomes, but none with a deletion of a group A chromosome. Van den Berghe et al. (5) described a newborn baby with gross abnormalities and a partial trisomy 1q, who died five hours after birth. It is possible that if a duplication of chromosome 1 causes such gross abnormalities, then a deficiency of this chromosome would be incompatible with life as the total loss of genetic material would be comparable to a monosomy D.

B. Group B

Duplications of chromosomes in this group are well documented, particularly in the case of trisomy 4p (4p+). Two of the earlier cases published as trisomy Bp (7, 40) are almost certainly 4p+ on clinical grounds, but the case of Shaw et al. (85) seems more likely to be 5p+. One case of 4p+ has been included because of the facial resemblance to other published cases, although no ocular abnormalities were described (52). In a review of eight cases, Réthoré and her coauthors (67) have established the complete phenotype of 4p+, in particular, a characteristic facial appearance with flat forehead, prominent glabella, hypoplasia of the nasal bones and a rounded tip of the nose, low-set ears, broadening of the concha and of the helix, a large tongue, a protruding chin, and a short neck (Fig. 10-1). In addition, these patients showed severe mental and physical retardation. The only ocular abnormality common to all is hypertelorism. A number of cases of trisomy 4p have been published since this review, and their facial resemblance is striking. There are also three instances in which siblings were also affected (32, 76, 77).

Trisomy 4p may be compared with the 4p- syndrome, in that the two syndromes have microcephaly, mental retardation, and skeletal malformation in common, but with several features which are the countertype of one another. These are compared in Table 10-2. It is hardly necessary to stress the advantages of the simple recognition of a syndrome such as trisomy 4p, when it arises out of a parental translocation and whereby several members of a family may be affected. There may also be a history of spontaneous abortion in the family due to a different type of segregation resulting in another unbalanced form of the translocation. There are only a few cases of 5p+ and also of 4q+ and 5q+, but not a sufficient number to establish a phenotype.

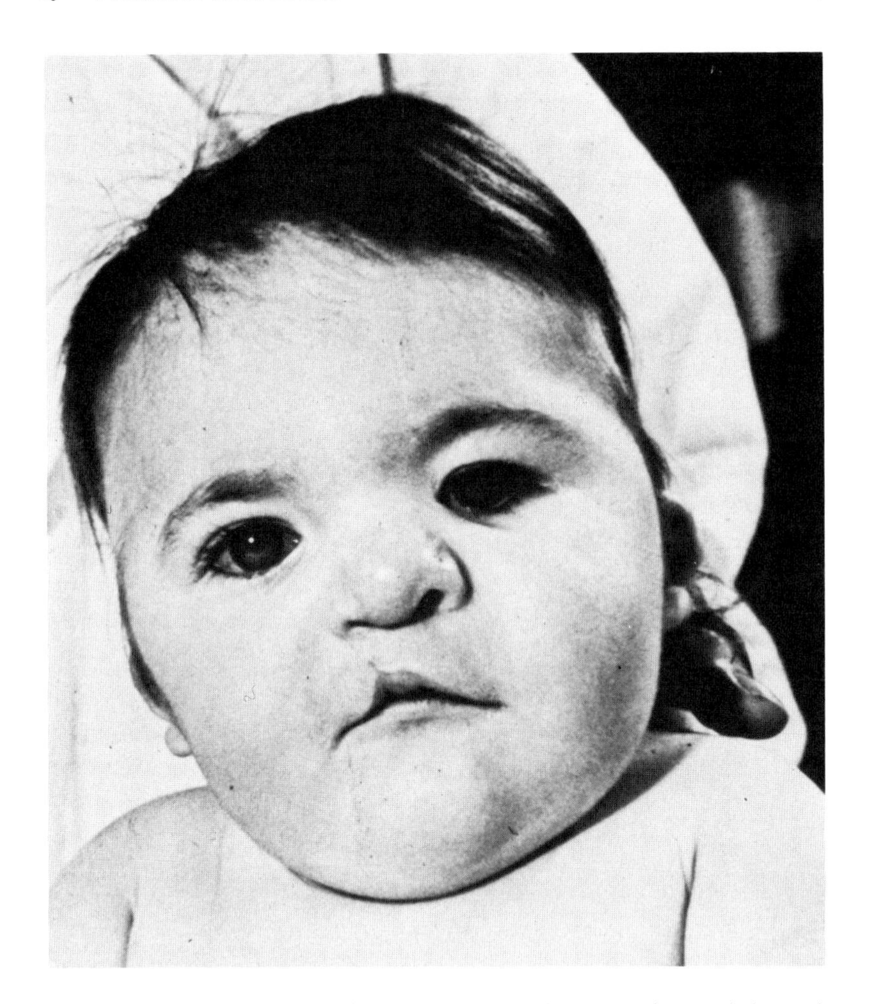

Fig. 10-1. 4p+ with facial asymmetry, right epicanthus and the right eye smaller and more deep-set than the left. (By courtesy of Dr. M.-O. Réthoré)

C. Group C

This group is numerically large, corresponding to the numerically large group of C chromosomes, but again, there is a conspicuous lack of reports of C deletions for purposes of comparison. However, among the C chromosomes, partial trisomy 9p has been studied by Réthoré and

and her coauthors (71) and the syndrome as described includes seven cases with moderate microcephaly, enophthalmos, antimongoloid slant, mild hypertelorism, protruding ears, globulous nose, downward slanting mouth, hypoplasia of the phalanges, and severe mental retardation (Figs. 10-2, 10-3). It is interesting that among the cases of 9p+, one

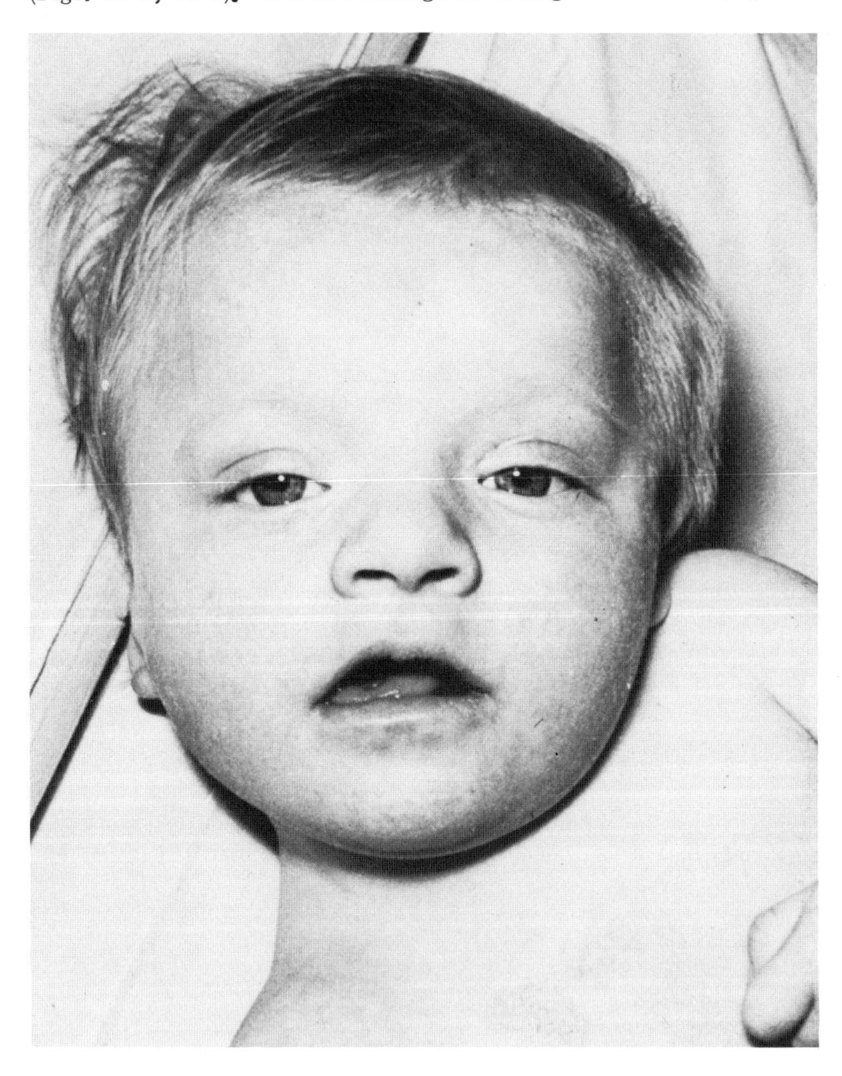

Fig. 10-2. 9p+ by t(9;21) with dolichocephaly, bulbous nose, and mild hypertelorism. (By courtesy of Dr. Renata Lax and Dr. Michael Ridler)

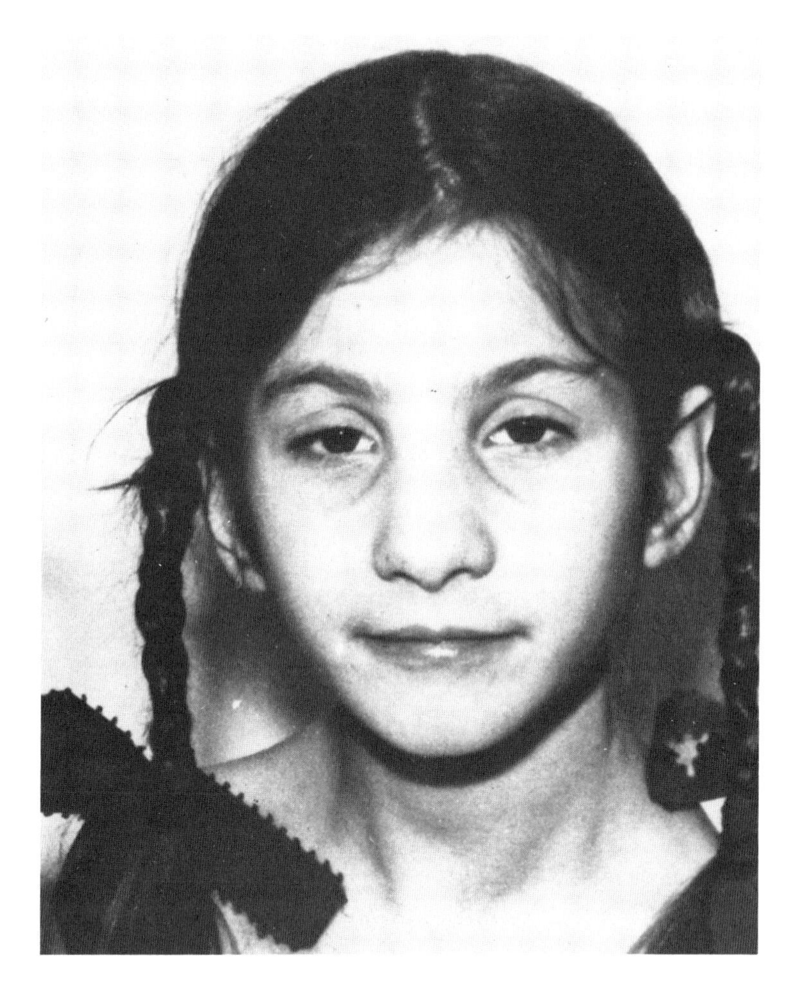

Fig. 10-3. 9p+. One of two affected sisters with deep-set eyes and mild hypertelorism. (By courtesy of Dr. Renata Lax and Dr. Michael Ridler)

(43) was not due to an inherited translocation although the father had an elongated paracentric long arm constriction of a C-group member. In the case reported by Réthoré et al. (70), the parents had normal karyotypes and the patient was 9p+ by a centric fusion with a t(15q9p) chromosome.

There are still too few cases of other partial trisomies in group C to establish other phenotypes or to compare them with trisomy C. This is either a question of time when reports on the subject will appear, or it may be that chromosome 9 is more fragile than the others in group C and more liable to breakage and, therefore, to translocations and inversions.

D. Group D

In group D there are a few cases of 13q+, but again not enough to make valid comparisons with 13q-. It would seem that clinically, they do have several features of trisomy 13, and this is only to be expected. Chromosome 13 is acrocentric with most of its material on the long arm, so that duplication of part or the whole of its long arm would produce signs similar to the duplication of the whole chromosome. One case of partial trisomy Dp is that described by Jacobsen et al. (47) with the mother and other members of her family as probable translocation carriers of a (DqDq) and a (DpDp) chromosome resembling a G type chromosome, where the proband inherited the DpDp chromosome and is therefore Dp+.

In this context it is worth mentioning the "cat-eye" syndrome, first described by Schachenmann and his associates (78), and since by several groups of authors (9, 15, 31, 61, 82, 91). It is characterized by iris coloboma, anal atresia, and mental retardation associated with an extra small metacentric or submetacentric chromosome. Other congenital abnormalities may be present such as preauricular fistulae and cardiac abnormalities, as well as hypertelorism and an antimongoloid slant. The coloboma may involve the choroid (15, 31, 61, 78), and the retina (15, 91). Alternating convergent strabismus was found in one case (91), and divergent strabismus with nystagmus in another (9). Ginsberg et al. (31) have described a case belonging to this syndrome. Their findings included anal atresia, a systolic cardiac murmur, uveal coloboma, cataract, retinal dysplasia, optic atrophy, and defective development of the angle of the anterior chamber. These ocular abnormalities are commonly found in trisomy 13, although this case also had numerous other ocular abnormalities which are not typical of trisomy 13, nor similar to others in the "cat-eye" syndrome. It may have been a partial trisomy 13, or perhaps a (DpDp) chromosome. This syndrome may occur in an incomplete form with anal atresia, mental retardation but no coloboma (6), or with no anal atresia (9). The origin of the small metacentric chromosome is somewhat obscure, and some authors suggest an isochromosome for the short arm of an acrocentric (Dp)i (61), a (17p)i or (18p)i (91), and a (21p)i (6), or a partial trisomy 22 with a 22q- chromosome (9). There are reports of "cat-eye" syndromes with normal karyotypes, but these have not been included here.

E. Group E

There are very few references of partial trisomies in the E group. One case is described by Hautschteck et al. (42) in which the mother had a D/E translocation, one of her children had many features of trisomy D with persistent hyaloid artery, although with no other ophthalmological abnormalities characteristic of trisomy D, and the other child had some characteristics of trisomy E with epicanthus, strabismus, and nystagmus. One child had duplication of part of chromosome D, while the other had duplication of 17p (Fig. 10-4).

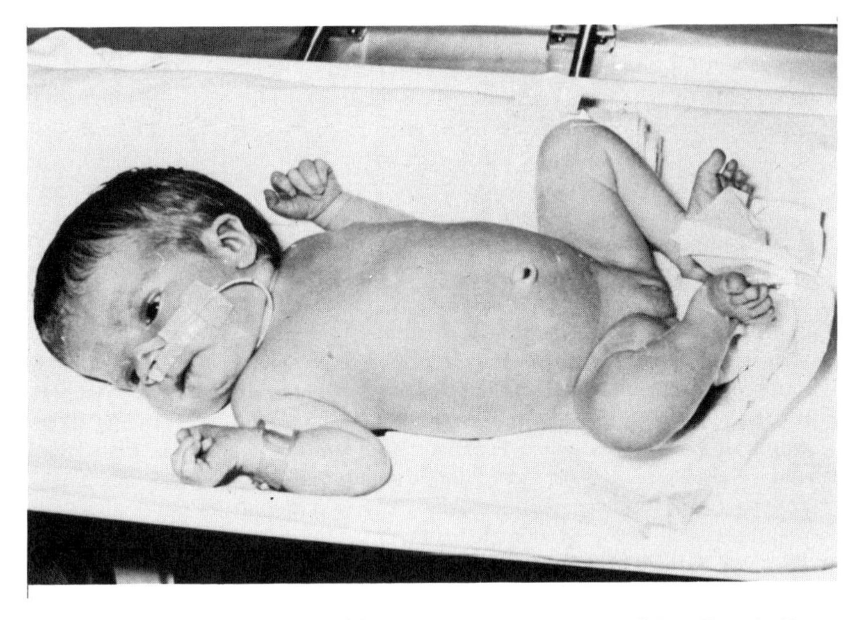

Fig. 10-4. 18q+ with frontal bossing. (By courtesy of Dr. Renata Lax and Dr. Michael Ridler)

F. Groups F and G

There are very few references in these two groups. Unfortunately, there is only one case of a partial trisomy 22 with a "cat-eye" syndrome (9), so that no comparison may be made with the corresponding deletion syndrome GII.

II. SUMMARY

A total of 98 cases of partial trisomy has been discussed. It is possible by using figures for the proportion of the total length, to calculate the expected number of cases in each chromosome group and to compare it with the number of cases reported. This is not intended to be a statistical review since there is a bias of selection and of ascertainment. Nearly all these cases are ascertained through a combination of mental or physical retardation, congenital malformations and spontaneous abortions, while very few are ascertained through general population surveys. There are however, one or two interesting aspects about the number of cases of trisomy 4p and 9p.

It would seem, for some reason, that both chromosomes 4 and 9 are liable to breakage and there is a relatively high incidence of translocations involving either of these two chromosomes. There have been studies on the evolution of the karyotype in man and the anthropoid apes. Man has fewer chromosomes than anthropoid apes, but relatively more metacentric and fewer acrocentric chromosomes. It is thought that the human metacentric chromosomes have evolved from Robertsonian translocation of acrocentric chromosomes in anthropoid apes. The human chromosomes have evolved further by pericentric inversions and reciprocal translocations. For instance, the gibbon, the orang-outan, the gorilla, and the chimpanzee have 48 chromosomes, with banding patterns similar to human chromosomes. The chimpanzee has 13 autosomes which are equivalent to human autosomes, but among those which are not equivalent are chromosomes 4 and 9 which in man have undergone a pericentric inversion. It is possible that these chromosomes may have retained a tendency to breakage at the original breakpoints, which would account for the relatively high incidence of reciprocal translocations involving these chromosomes. There are various theories about breakpoints in chromosomes, and the relative fragility of chromosomes or even of specific bands. There have been analyses of breakpoints in translocations and inversions which suggest that these do not occur at random.

While chromosomes 4 and 9 seem to behave in a special manner, there is one peculiarity of chromosome 9 which should be mentioned. Chromosome 9 has a paracentric secondary constriction on the long arm which is heterochromatic and varies in size among the population. This region is known as 9qh, and its variation in size produces no apparent phenotypic effect and it is a common variant in the population.

Another interesting feature of partial trisomies is that in this review, in only seven cases out of 98 were the karyotypes of the parents normal. Furthermore, it seems that there is no significant difference between the numbers of maternal and paternal translocations. It has been said previously that in a D/G translocation carrier, mothers are given a one in six risk of giving birth to offspring with trisomy 21, and that the risk for fathers with a D/G translocation is much less. In view of the small number of cases reported with partial trisomy, it is too early to speculate on the relatively high number of paternal translocations involved, although the explanation will be of great importance in genetic counselling.

TABLE 10-1

Partial Trisomy Transmitted by Parental Translocation

Author	Ocular Abnormalities	Partial Trisomy	Parental Translocation
van den Berghe et al. (5) 1973	Eyes small and wide set; fleshy mass with wrinkled skin and abundant hair-growth on the forehead, above left eye	1q	1q-;12p+mat
Norwood and Hoehn (62) 1974	Antimongoloid slant	1q	1;3pat
Bender et al. (4) 1969	Hypertelorism, antimongoloid slant, and convergent strabismus	2p	2p-;Cp+pat
Forabosco et al. (25) 1973	Frontal bossing, hypertelorism, antimongoloid slant, divergent strabismus, pendular nystagmus	2q	2q-;13q+mat
Clarke et al. (16) 1964	Alternating strabismus	3?	3?-;Cq+mat
Walzer et al. (98) 1966	Bilateral coloboma, left lower lid hypoplastic, left orbit shallow; five other affected children in family; all parents were carriers	3?	3?-;Bp+mat
Aarskog (1) 1969	Hypertelorism, epicanthus, very pale optic discs	3?	3?+;18q-mat
Réthoré et al. (72) 1972	Marked hypertelorism, bilateral epicanthus	3p	3p-;Cq+mat
Böök et al. (7) 1963	Marked hydrocephalus, coloboma of irides	Bp	Bp-;Gq+mat

TABLE 10-1 (continued)

Author	Ocular Abnormalities	Partial Trisomy	Parental Translocation
Gustavson et al. (40) 1964	Slight hypertelorism, very small palpebral fissures, blepharophimosis, and bilateral iris coloboma	Bp	Bp-;Gp+mat
Shaw et al. (85) 1965	Flattened head, right eye protuberant with grey discolouration of the globe, vascularized cornea, patches of opacity; left eye sunken and less than half normal size	Bp	Bq-;Bp+mat
Wilson et al. (100) 1970	Right eye normal; left eye, keyhole coloboma of the iris, retina, and choroid	4p	maternal inv(4)
Gouw et al. (34) 1972	Marked asymmetry of the skull; hypertelorism	4p	4p-;14+ or 15+pat
Schinzel and Schmid (79) 1972	Hypertelorism, narrow palpebral fissures	4p	4p-;18q+pat
Metz et al. (56) 1973	No ophthalmological abnormalities	4p	4p-;22p+mat
Schwanitz and Grosse (83) 1973	Left pupil slightly larger than the right	4p	4p-;22p+pat
Giovannelli et al. (32) 1974a	Case 1: hyperplastic supraorbital ridges, slight antimongoloid slant, hypertelorism, protuberant main branch of the central vein of the pupil; her sister, case 2, same supraorbital ridges, pendular nystagmus, normal fundi	4p	4p-;22p+mat

TABLE 10-1 (continued)

Author	Ocular Abnormalities	Partial Trisomy	Parental Translocation
Giovannelli et al. (33) 1974b	Further description of the two cases above	4p	4p-;22p+mat
Réthoré et al. (68) 1974b	Right epicanthus with the eye smaller and more deep-set than the left, bilateral nystagmus, fundus mottled ("granite")	4p	inv4(p14q35) pat
Dallapiccola et al. (18) 1974	Bushy eyebrows meeting over the nasal bridge, hypertelorism, antimongoloid slant, right convergent strabismus	4p	inv4(p13q35) pat
Owen et al. (63) 1974	Narrow palpebral fissures, bilateral microphthalmos, slight antimongoloid slant	4p	4p-;22p+pat
Sartori et al. (77) 1974	2 sibs: female with bushy eyebrows, long lashes, divergent strabismus, nystagmoid jerks on extreme lateralized gaze; brother with very similar features	4p	4p-;22q+pat
Furbetta et al. (29) 1975	Depressed and broad nasal bridge, bushy eyebrows, long eyelashes, narrow palpebral fissures, hypertelorism	4p	4p-;21q+mat
Edwards et al. (22) 1962	Case 1: hypertelorism, slight epicanthus; case 2: a sib, had white, speckled and hypoplastic irides	4q	9p-;4q+pat
Surana and Conen (90) 1972	Hypertelorism	4q	4q-;18q+mat
de la Chapelle et al. (14) 1973	Bird face, slanting forehead, narrow palpebral fissures, slight epicanthic folds	4q	4q-;21q+mat

TABLE 10-1 (continued)

Author	Ocular Abnormalities	Partial Trisomy	Parental Translocation
Fonatsch et al. (24) 1974	Mild midface retraction, deep-set eyes, epicanthus, hypertelorism, convergent strabismus, bilateral hyperopia of 3D, normal fundi	4q	4q-;18q+pat
Revazov et al. (73) 1971	Moon face, hypertelorism, microphthalmos, antimongoloid slant, bilateral epicanthus, iris coloboma, convergent strabismus, ptosis	Bq	Not tested
Knight et al. (48) 1971	Bilateral epicanthus, ptosis	5q	5q-;14p+mat
Gray et al. (36) 1966a	Small pupils with poor reaction to light, probable blocked tear ducts	Cp	Probable mat(GqCp)
Lord et al. (54) 1967	Case 1: "nodules" over left eyelid	Cp	? Cp-;Gq+mat
Eriksson et al. (23) 1968	Asymmetrical face with large eyes	Cp mosaic	Parents normal
Deminatti et al. (19) 1969	Slight bilateral epicanthus, divergent strabismus	Cp	Cp-;Gp+mat
Cantu et al. (12) 1971	Case 1: marked hypertelorism, antimongoloid slant; case 2 is a brother of case 1 with the same ocular abnormalities	Cp	Parents normal
Yanagisawa and Hiraoka (101) 1971	Case 1: strabismus. A brother, case 2: hypertelorism. A second cousin, case 3: cataract in left eye, microphthalmos right eye	Cp	Cp-;Gp+ carriers

TABLE 10-1 (continued)

Author	Ocular Abnormalities	Partial Trisomy	Parental Translocation
Lindsten et al. (53) 1965	Slight hypertelorism	? Cq	Cq-; Cp+pat
Gray et al. (37) 1966b	Moderate hypertelorism	Cq	Cp-; Gp+mat
Mikelsaar et al. (58) 1967	Hypertelorism, bilateral epicanthus, one eye lower than the other, strabismus	Cq	Cq-; Bq+pat
Mikkelsen et al. (59) 1968	Slight bilateral exophthalmos, right divergent strabismus	Cq	Cq-; Gq+pat
Warburg and Andersen (99) 1968	Detailed study of Mikkelsen (59): slight peripheral iris atrophy, an anterior synechia to the peripheral cornea, iridoschisis	Cq	Cq-; Gq+pat
Rott et al. (75) 1971	Mongoloid slant	Cq	4p+; 9q-mat
Talvik et al. (92) 1973	Microphthalmos, narrow palpebral fissures, slight antimongoloid slant, hypertelorism	Cq (10 or 11q)	Cq-; Dq+pat (10/11q-; 14q+)
Mulcahy et al. (60) 1974	Short palpebral fissures which could not be opened	Cq	t(10;13)(q24 q34)mat
Rhode and Catz (74) 1964	Ectopic pupils, myopia and astigmatism, exotropia right eye	6q	6q-; 9q+mat
Alfi et al. (2) 1973	Asymmetric head with prominent eyes	7q	7q-; 14q+pat
Grace et al. (35) 1973	Epicanthus, almond-shaped eyes, rotatory nystagmus, left concomitant convergent strabismus	7q	7q-; 3q+mat
Vogel et al. (96) 1973	Hypertelorism, iris coloboma	7q	7q-; 21p+mat

TABLE 10-1 (continued)

Author	Ocular Abnormalities	Partial Trisomy	Parental Translocation
Lozzio and Kattine (55) 1969	Mongoloid slant, haemorrhagic sclerae	8p	Cp-;2q+pat
Fryns et al. (26) 1974	Marked hypertelorism, exophthalmos, ptosis, telangiectases in sclera	8q	parents normal
Butler et al. (10) 1974	Narrow palpebral fissures, slight microphthalmos	9p	Dq-;Cp+mat
Réthoré et al. (71) 1970	Case 1: small deep-set eyes, convergent strabismus;	9p	9q-;6p+mat
	case 2: large asymmetrical palpebral fissures, hypertelorism, epicanthus, convergent strabismus;	9p	9q-;Gq+mat
	case 3: deep-set eyes, antimongoloid slant, convergent strabismus, slightly ectopic pupils;	9p	9p-;Gp+pat
	case 4 (brother of case 3): almost completely blind, deep-set eyes, antimongoloid slant, iris colobomas, retinal detachment	9p	9p-;Gp+pat
Hoehn et al. (43) 1971	Small deep-set eyes, antimongoloid slant, moderate hypertelorism, convergent strabismus	9p	parents normal
Baccichetti and Tenconi (3) 1973	Antimongoloid slant, hypertelorism, convergent strabismus	9p	parents normal
Réthoré et al. (69) 1973	Deep-set eyes, hypertelorism, antimongoloid slant	9p by t(15q9p)	parents normal
Fujita et al. (28) 1974	Epicanthus, hypertelorism, broad nasal bridge, normal fundi	9p	complex mat. translocation

TABLE 10-1 (continued)

Author	Ocular Abnormalities	Partial Trisomy	Parental Translocation
Réthoré et al. (70) 1974c	Antimongoloid slant, enophthalmos	9p	4q-;9q+mat
Schwanitz et al. (84) 1974	Hypertelorism, mongoloid slant, convergent strabismus, right coloboma	9p	t(8;9)(q24q22) mat
Turleau et al. (93) 1974	Case 1: deep orbits, eyes slightly almond-shaped, bilateral epicanthus;	9p	9q-;15q+mat
	case 2: microcephaly, frontal bossing, slight hypertelorism, mongoloid slant, Brushfield spots	9p	9p-;7q+mat
Zaremba et al. (103) 1974	Case V_1: hypertelorism, antimongoloid slant, hypermetropic astigmatism, strabismus; case V_3: hypertelorism, hypermetropia, strabismus; case IV_{14}: hypertelorism, myopic astigmatism, strabismus; case IV_{16}: myopic astigmatism, hypertelorism	9p	t(9;15)(q13;q11) all carriers
de Grouchy and Canet (38) 1965	Slight antimongoloid slant, hypertelorism, microcornea (left eye), small atrophy of pigment epithelium, very tortuous retinal vessels evoking angiomatosis	10q	10q-;Dq+mat
Dutrillaux et al. (21) 1973b	Microphthalmos with narrow palpebral fissures, hypertelorism, no papillary oedema, veins seem dilated	10q	inv(10)(p15q 24)mat

TABLE 10-1 (continued)

Author	Ocular Abnormalities	Partial Trisomy	Parental Translocation
Laurent et al. (50) 1973	Antimongoloid slant, hypertelorism, bilateral microphthalmos, very pale papillae (? optic atrophy), palpebral phimosis	10q	t(1;10)(q44; q22)mat
Talvik et al. (92) 1973	Proband of family 1: microcephaly, microphthalmos, narrow palpebral fissures, antimongoloid slant, hypertelorism, broad nasal bridge, prominent glabella, weak pupillary reaction to light, optic papillae light pink	10q	14q+;10q-pat
Yunis and Sanchez (102) 1974	Small palpebral fissures, antimongoloid slant, ptosis, microphthalmos, bilateral lens opacity, absence of anterior chamber, retina replaced by fibrous tissue, pendular nystagmus	10q	pat trans
Hustinx et al. (46) 1974	Epicanthus, hypertelorism, broad nasal bridge	10p	10p-;14p+mat
Rott (76) 1972	Two children in a family with hypertelorism, and epicanthus	11q	11q-;13q+ carriers
Lejeune et al. (51) 1966	Marked hypertelorism, epicanthus, antimongoloid slant	12p	12p-;Fmat
Uchida and Lin (94) 1973	Prominent epicanthus, Brushfield spots	12p	12p-;8p+pat
Holboth et al. (44) 1974	Flattened head, hypertelorism	12q	12q-;21p+mat

TABLE 10-1 (continued)

Author	Ocular Abnormalities	Partial Trisomy	Parental Translocation
Stalder et al. (87) 1964	Convergent strabismus, eyes otherwise normal	Dq	Dq-;Cp+pat
Hautschteck et al. (42) 1966	Case 1: hypertelorism, persistent hyaloid artery bilaterally; case 2 (sib of case 1): epicanthus, convergent strabismus, and intermittent nystagmus	D?+ distal D?+prox and 17p+ distal	presumptive mat trans
Jacobsen et al. (47) 1966	Epicanthus. Left eye: iris coloboma, lens clear except in the area of coloboma where there are two cortical opacities, small coloboma of the lens	Dp	mat ?DqDq and DpDp
Hsu et al. (45) 1973	Bilateral iris colobomas, capillary hemangioma of the occipital region	13q by t(13;13), (p12;q13)	Parent normal
Talvik et al. (92) 1973	Proband of family 2: mongoloid slant	13q	13q-;21q+pat
Schinzel et al. (81) 1974b	Case 2: hypertelorism, long and incurved eyelashes, strabismus, myopia and astigmatism; case 3 (brother of case 2): epicanthus, nystagmus, strabismus, choroidal atrophy, pale optic discs, myopia and astigmatism	13q 13q	t(13;17) (q14;p13)mat
Pfeiffer et al. (64) 1973	Antimongoloid slant, buphthalmos	14q	14q-;21q+pat
Fujimoto et al. (27) 1974	Microcephaly, antimongoloid slant, central cataracts, esotropia	15q	Dq-;Gq+mat

TABLE 10-1 (continued)

Author	Ocular Abnormalities	Partial Trisomy	Parental Translocation
Latta and Hoo (49) 1974	Hypertelorism, ptosis	17p	Parents normal
Valdmanis et al. (95) 1967	Proband, see under 18q-; paternal aunt had divergent strabismus	18q	18q-;4p+pat
Steele et al. (88) 1974	Hypertelorism, marked convergent strabismus	18q	18q-;4q+pat
Carrel et al. (13) 1971	Three siblings: hypertelorism, 3/3; mongoloid slant, 2/3; iris coloboma, 1/3	Fp	Fp-;13q+mat
Šubrt and Brychnáč (89) 1974	Prominent bilateral epicanthus	20p	20p-;21p+mat
Dutrillaux et al. (20) 1973a	Mentally retarded female aged 54, congenital microphthalmos, blind	Unbalanced	4q/21q

TABLE 10-2

Type and Countertype of 4p- and 4p+ *

Symptoms in Common	Feature	Countertype Symptoms 4p-	4p+
Microcephaly	Forehead	Rounded	Flat
	Glabella	Aplastic	Prominent
Skeletal malformations	Nasal bones	Prominent	Hypoplastic
	Tip of nose	Angular	Spherical
Mental retardation	Chin	Receding	Pointed
	Neck	Long	Short

* Modified from Réthoré et al. (67)

III. PERICENTRIC INVERSIONS

The ocular abnormalities associated with pericentric inversions are listed in Table 10-3. There are too few references to draw any significant conclusions about the effect of an inversion on the phenotype, but the usual signs associated with some chromosomal imbalance are present. It will be noted that in one family, a number of cases of duplication resulted from the transmission of a parental inversion where the carriers were phenotypically normal. This is due to crossing over at meiosis between the inverted chromosome and its normal homologue, which leads to duplication and deficiency in the recombinant chromosomes (41). There are also cases of inversions where the parents are phenotypically normal and who may have either a normal karyotype or carry the same inversion. The discussion of these cases is very much the same as in the next section on balanced translocations.

No references were found for paracentric inversions. This does not imply that they do not occur, but that their detection is more difficult as the inverted chromosome remains morphologically identical to its homologue.

TABLE 10-3

Pericentric Inversions

Author	Ocular Abnormalities	Chromosome Inversion	Parental Inversion
Lele et al. (52) 1965	Case 1: coloboma of iris of one eye; case 3: bilateral colobomas of the upper and inner quadrant of the iris	inv(1) inv(1)	mat inv parents normal
Mikelsaar et al. (57) 1970	Narrow palpebral fissures, slight antimongoloid slant, right microphthalmos, left iris coloboma	1p+q-	Parents normal
de Grouchy et al. (39) 1963	Hypertelorism, slight mongoloid slant, myopia, and astigmatism	inv(2)	Parents not studied
Podugolnikova and Blumina (66) 1972	Hypertelorism, slight epicanthus, and mongoloid slant	inv(2)	mat inv(2)
Cohen and Davidson (17) 1971	Bilateral inferior colobomas of the irides	3p-q+	mat inv(3)
Boué et al. (8) 1974	Flat forehead, hypoplastic supraorbital ridges, mongoloid slant, slight hypertelorism	3p-q+	inv(3)(p25 q21)pat
Butler et al. (11) 1974	Convergent strabismus	3p-q+	
Soukup et al. (86) 1974	Low forehead, prominent supraorbital ridges, slight synophris, blepharophimosis, antimongoloid slant	inv(4)	t(13q14q)
Pitt et al. (65) 1967	Divergent strabismus, hypertelorism, epicanthus	Cp+q-	Parents normal

TABLE 10-3 (continued)

Author	Ocular Abnormalities	Chromosome Inversion	Parental Inversion
Wahrman et al. (97) 1972	Proband of family 2: buphthalmos	9p+q-	pat inv
Schinzel et al. (80) 1974a	Hypoplastic supraorbital ridges, hypertelorism, mongoloid slant, blepharophimosis, deeply-set eyes of normal size	inv(9) mosaic	inv(9)mat
Hauksdóttir et al. (41) 1972	Study of a large family: two of the affected subjects had hypertrichosis of the lashes, strabismus, and one had optic atrophy	Duplication of 13q and deficiency of 13p by inv(13) in carriers	

REFERENCES

1. Aarskog, D. A familial 3/18 reciprocal translocation resulting in chromosome duplication deficiency (3?+:18q-). Acta Paediatr. Scand. 58: 397-406 (1969).

2. Alfi, O. S., Donnell, G. N. and Kramer, S. L. Partial trisomy of the long arm of chromosome No. 7. J. Med. Genet. 10: 187-189 (1973).

3. Baccichetti, C. and Tenconi, R. A new case of trisomy for the short arm of No 9 chromosome. J. Med. Genet. 10: 296-299 (1973).

4. Bender, K., Reinwein, H., Gorman, L. Z. and Wolf, U. Familiäre 2/C Translokation: 46,XY,t(2p-:Cp+) und 46,XX, Cp+. Humangenetik 8: 94-104 (1969).

5. Berghe, H. van den, Eygen, M. van, Fryns, J. P., Tanghe, W. and Verresen, H. Partial trisomy 1, karyotype 46,XY,12-,t(1q-;12p+). Humangenetik 18: 225-230 (1973).

6. Beyer, P., Ruch, J. -V., Rumpler, Y. and Girard, J. Observation d'un enfant débile mental et polymalformé dont le caryotype montre la présence d'un petit extra-chromosome médiocentrique. Pediatrie (Lyon) 23: 439-442 (1968).

7. Böök, J. A., Atkins, L. and Santesson, B. Some new data on autosomal aberrations in man. Pathol. Biol. (Paris) 11: 1159-1162 (1963).

8. Boué, J., Hirschhorn, K., Lucas, M., Gautier, M., Moszer, M. and Bach, Ch. Aneusomies de recombinaison. Conséquences d'une inversion péricentrique d'un chromosome 3 paternel. Ann. Pediatr. (Paris) 21: 567-573 (1974).

9. Bühler, E. M., Méhes, K., Müller, H. and Stalder, G. R. Cat-eye syndrome: A partial trisomy 22. Humangenetik 15: 150-162 (1972).

10. Butler, L. J., Eades, S. M. and France, N. E. Transmission of a translocation t(Cp+:Dq-) through three generations including an example of probable trisomy for the short arm of the C group chromosome No 9. Ann. Genet. (Paris) 12: 15-27 (1969).

11. Butler, L. J., Hall, M. E. and Wharton, B. A. A retarded child with a 46,XX,3p-q+ chromosome karyotype. J. Ment. Defic. Res. 18: 41-49 (1974).

12. Cantu, J. -M., Buentello, L. and Armendares, S. Trisomie Cp: Un nouveau syndrome. Ann. Genet. (Paris) 14: 177-186 (1971).

13. Carrel, R. E., Sparkes, R. S. and Wright, S. W. Partial F trisomy with familial F/13 translocation detected and identified by parental chromosome studies. J. Pediatr. 78: 664-672 (1971).

14. Chapelle, A. de la, Koivisto, M. and Schröder, J. Segregating reciprocal (4:21)(q21;q21) translocation with preposita trisomic for parts of 4q and 21. J. Med. Genet. 10: 384-389 (1973).

15. Chieri, R. de., Malfatti, C., Stanchi, F. and Albores, J. M. Cat-eye syndrome: Evaluation of the extra chromosome with banding techniques. Case report. J. Genet. Hum. 22: 101-107 (1974).

16. Clarke, G., Stevenson, A. C., Davies, P., Williams, C. E. and Holt, S. B. A family apparently showing transmission of a translocation between chromosome 3 and one of the "X-6-12" or "C" group. J. Med. Genet. 1: 27-32 (1964).

17. Cohen, M. M. and Davidson, R. G. An inherited pericentric chromosomal inversion (46, inv3 p-q+) associated with skeletal anomalies. J. Pediatr. 79: 456-462 (1971).

18. Dallapiccola, B., Capra, L., Preto, G., Covic, M. and Dutrillaux, B. Inversion péricentrique du 4:inv(4)(p13q35) et trisomie du bras court du 4 par aneusomie de recombinaison. Ann. Genet. (Paris) 17: 115-118 (1974).

19. Deminatti, M., Maillard, E., Gosselin, B., Peltier, J. M., Bulteel, M. F. and Dupuis, C. Trisomie partielle C par translocation t(Cp-:Gp+). Ann. Genet. (Paris) 12: 36-45 (1969).

20. Dutrillaux, B., Jonasson, J., Laurén, K., Lejeune, J., Lindsten, J., Petersen, G. B. and Saldaña-Garcia, P. An unbalanced 4q/21q translocation identified by the R but not by the G and Q chromosome banding techniques. Ann. Genet. (Paris) 16: 11-16 (1973a).

21. Dutrillaux, B., Laurent, C., Robert, J. M. and Lejeune, J. Inversion péricentrique, inv(10), chez la mère et aneusomie de recombinaison, inv(10), rec(10), chez son fils. Cytogenetics (Basel) 12: 245-253 (1973b).

22. Edwards, J. H., Fraccaro, M., Davies, P. and Young, R. B. Structural heterozygosis in man: Analysis of two families. Ann. Hum. Genet. 26: 163-169 (1962).

23. Eriksson, B., Fraccaro, M., Hulten, M., Lindsten, J. and Tiepolo, L. Unusual chromosomal mosaic (46, XX/46, XXCp+) in a girl with multiple malformations. Ann. Genet. (Paris) 11: 6-10 (1968).

24. Fonatsch, C., Flatz, S. D. and Hürter, P. Partial trisomy 4q and partial monosomy 18q as a consequence of a paternal balanced translocation. Humangenetik 25: 227-233 (1974).

25. Forabosco, A., Dutrillaux, B., Toni, G., Tamborino, G. and Cavazzuti, G. Translocation équilibrée t(2:13)(q32:q33) familiale et trisomie 2q partielle. Ann. Genet. (Paris) 16: 255-258 (1973).

26. Fryns, J. P., Verresen, H., Berghe, H. van den, Kerckvoorde, J. van and Cassiman, J. Partial trisomy 8: Trisomy of the distal part of the long arm of chromosome number 8 +(8q2) in a severely retarded and malformed girl. Humangenetik 24: 241-246 (1974).

27. Fujimoto, A., Towner, J. W., Ebbin, A. J., Kahlstrom, E. J. and Wilson, M. G. Inherited partial duplication of chromosome No. 15. J. Med. Genet. 11: 287-290 (1974).

28. Fujita, H., Abe, T., Yamamoto, K. and Furuyama, J. Possible complex translocation t(9;14;13)(q12;p1?;q31) in mother of a child with 9p- trisomy syndrome. Humangenetik 25: 83-92 (1974).

29. Furbetta, M., Rosi, G., Cossu, P. and Cao, A. A case of trisomy of the short arms of chromosome No. 4 with translocation t(4p21p;4q21q) in the mother. Humangenetik 26: 87-91 (1975).

30. Gemme, G., Vianello, M. G., Zera, M. and Reboa, E. Un caso di sindrome polimalformativa con traslocazione cromosomica t(8q+;9q-). Minerva Pediatr. 25: 1338-1344 (1973).

31. Ginsberg, J., Dignan, P. and Soukup, S. Ocular abnormality associated with extra small autosome. Am. J. Ophthalmol. 65: 740-746 (1968).

32. Giovannelli, G., Forabosco, A. and Dutrillaux, B. Translocation familiale t(4:22)(p11:p12) et trisomie 4p chez deux germains. Ann. Genet. (Paris) 17: 119-124 (1974a).

33. Giovannelli, G., Pezzani, C., Medioli Cavara, F. Indagine neuropsicopatologica in 2 casi di trisomia 4p. Minerva Pediatr. 26: 1933-1943 (1974b).

34. Gouw, W. L., Anders, G. J. P. A., Ten Kate, L. P. and Groot, C. J. de. Paternal transmission of a B/D translocation t(4p-:14p+or15p+) resulting in a partial 4p trisomy. Humangenetik 16: 251-259 (1972).

35. Grace, E., Sutherland, G. R., Stark, G. D. and Bain, A. D. Partial trisomy of 7q resulting from a familial translocation. Ann. Genet. (Paris) 16: 51-54 (1973).

36. Gray, J. E., Dartnall, J. A., Creery, R. D. G. and Croudace, J. Congenital anomalies due to transmission of a chromosome translocation. J. Med. Genet. 3: 59-61 (1966a).

37. Gray, J. E., Dartnall, J. A. and MacNamara, B. G. P. A family showing transmission of a translocation between a 6-12 chromosome and a 21-22 chromosome. J. Med. Genet. 3: 62-65 (1966b).

38. Grouchy, J. de and Canet, J. Translocation 6-12-13-15 et trisomie partielle 6-12 (probablement 10). Ann. Genet. (Paris) 8: 16-20 (1965).

39. Grouchy, J. de, Emerit, I., Corone, P., Vernant, P., Lamy, M. and Soulie, P. Inversion péricentrique probable du chromosome No 2 et malformations congénitales chez un garçon. Ann. Genet. (Paris) 6: 21-23 (1963).

40. Gustavson, K. -H., Finley, S. C., Finley, W. H. and Jalling, B. A 4-5/21-22 chromosomal translocation associated with multiple congenital anomalies. Acta Paediatr. Scand. 53: 172-181 (1964).

41. Hauksdóttir, H., Halldórsson, S., Jensson, O., Mikkelsen, M. and McDermott, A. Pericentric inversion of chromosome No 13 in a large family leading to duplication deficiency causing congenital malformations in three individuals. J. Med. Genet. 9: 413-421 (1972).

42. Hautschteck, E., Mursey, G., Prader, A. and Bühler, E. Siblings with different types of chromosomal aberrations due to D-E translocation of the mother. Cytogenetics (Basel) 5: 281-294 (1966).

43. Hoehn, H., Engel, W. and Reinwein, H. Presumed trisomy for the short arm of chromosome No 9 not due to inherited translocation. Humangenetik 12: 175-181 (1971).

44. Holboth, N., Jacobsen, P. and Mikkelsen, M. Partial trisomy 12 in a mentally retarded boy and translocation (12:21) in his mother. J. Med. Genet. 11: 299-303 (1974).

45. Hsu, L. Y. F., Kim, H. J., Sujansky, E., Kouseff, B. and Hirschhorn, K. Recriprocal translocation versus centric fusion between two No 13 chromosomes. A case of 46, XX, -13, +t(13:13) (p12:q13) and a case of 46, XY, -13, +t(13:13)(p12p12). Cytogenetics (Basel) 12: 235-244 (1973).

46. Hustinx, Th. W. J., Ter Haar, B. G. A., Scheres, J. M. J. C. and Rutten, F. J. Trisomy for the short arm of chromosome No 10. Clinical Genetics 6: 408-415 (1974).

47. Jacobsen, P., Mikkelsen, M., Frøland, A. and Dupont, A. Familial transmission of a translocation between two non-homologous large acrocentric chromosomes. Clinical, cytogenetic and autoradiographic studies. Ann. Hum. Genet. 29: 391-399 (1966).

48. Knight, L. A., Sakaguchi, S. and Luzzatti, L. Unusual mechanism of transmission of a maternal chromosome translocation. Am. J. Dis. Child. 121: 162-167 (1971).

49. Latta, E. and Hoo, J. J. Trisomy of the short arm of chromosome 17. Humangenetik 23: 213-217 (1974).

50. Laurent, C., Bovier-Lapierre, M. and Dutrillaux, B. Trisomie 10 partielle par translocation familiale t(1:10)(q44:q22). Humangenetik 18: 321-328 (1973).

51. Lejeune, J., Berger, R., Réthoré, M. -O., Salmon, Ch. and Kaplan, M. Translocation Cc~F familiale déterminant une trisomie pour le bras court du chromosome 12. Ann. Genet. (Paris) 9: 12-18 (1966).

52. Lele, K. P., Dent, T. and Delhanty, J. D. A. Chromosome studies in five cases of coloboma of the iris. Lancet 1: 576-578 (1965).

53. Lindsten, J., Fraccaro, M., Klinger, H. P. and Zetterqvist, P. Meiotic and mitotic studies of a familial recriprocal translocation between two autosomes of group 6-12. Cytogenetics (Basel) 4: 45-64 (1965).

54. Lord, P. M., Casey, M. D. and Laurence, B. M. A new translocation between chromosomes in the 6-12 and 21-22 groups. J. Med. Genet. 4: 169-176 (1967).

55. Lozzio, C. B. and Kattine, A. A. Familial transmission of a chromosome translocation t(2q+:Cp-). J. Med. Genet. 6: 174-179 (1969).

56. Metz, F., Bier, L. and Pfeiffer, R. A. Partielle Trisomie des kurzen Arms eines Chromosoms Nr 4 in der Folge einer Translokation t(4p-:22p+). Humangenetik 18: 207-211 (1973).

57. Mikelsaar, A. -V. N., Ananjev, E. V. and Gindilis, V. M. Probable pericentric inversion in chromosome No 1 in a female child (46, XX, inv(1p+q-)). Humangenetik 9: 316-324 (1970).

58. Mikelsaar, A. -V. N., Talvik, T. A. and Sitska, M. E. A presumable familial B/C translocation in man. (A preliminary report). Genetika 3(4): 146-152 (1967).

59. Mikkelsen, M., Mortensen, E., Skakkebaek, N. E. and Yssing, M. Familial reciprocal translocation between a C group (12?) chromosome and a late labelling G chromosome. Acta Genet. (Basel) 18: 241-250 (1968).

60. Mulcahy, M. T., Jenkyn, J. and Masters, P. L. A familial 10/13 translocation: Partial trisomy C in an infant associated with familial 10/13 translocation. Clinical Genetics 6: 335-340 (1974).

61. Noël, B. and Quack, B. A small submetacentric supernumerary chromosome in multiple deformity. J. Genet. Hum. 18: 45-55 (1970).

62. Norwood, T. H. and Hoehn, H. Trisomy of the long arm of human chromosome 1. Humangenetik 25: 79-82 (1974).

63. Owen, L., Martin, B., Blank, C. E. and Harris, F. Multiple congenital defects associated with trisomy for the short arm of chromosome 4. J. Med. Genet. 11: 291-295 (1974).

64. Pfeiffer, R. A., Büttinghaus, K. and Struck, H. Partial trisomy 14 following a balanced reciprocal translocation t(14q-:21q+). Humangenetik 20: 187-189 (1973).

65. Pitt, D. B., Weiner, S., Sutherland, G. and Pearce, P. The pericentric syndrome. Lancet 2: 568 (1967).

66. Podugolnikova, O. A. and Blumina, M. G. Pericentric inversion of the chromosome 2 in a girl with a congenital mental deficiency and in her mother. Genetika 8(8): 156-161 (1972).

67. Réthoré, M. -O., Dutrillaux, B., Giovannelli, G., Forabosco, A., Dallapiccola, B. and Lejeune, J. La trisomie 4p. Ann. Genet. (Paris) 17: 125-128 (1974a).

68. Réthoré, M. -O. Dutrillaux, B., Job, J. -C. and Lejeune, J. Trisomie 4p par aneusomie de recombinaison d'une inv(4)(p14q35). Ann. Genet. (Paris) 17: 109-114 (1974b).

69. Réthoré, M. -O., Hoehn, H., Rott, H. D., Couturier, J., Dutrillaux, B. and Lejeune, J. Analyse de la trisomie 9p par dénaturation ménagée. Humangenetik 18: 129-138 (1973).

70. Réthoré, M. -O., Ferrand, J., Dutrillaux, B. and Lejeune, J. Trisomie 9p par t(4:9)(q43:q21) mat. Ann. Genet. (Paris) 17: 157-161 (1974c).

71. Réthoré, M. -O., Larget-Piet, L., Abonyi, D., Boeswillwald, M., Berger, R., Carpentier, S., Gruveiller, J., Dutrillaux, B., Lafourcade, J., Penneau, M. and Lejeune, J. Sur quatre cas de trisomie pour le bras court du chromosome 9. Individualisation d'une nouvelle entité morbide. Ann. Genet. (Paris) 13: 217-232 (1970).

72. Réthoré, M. -O., Lejeune, J., Carpentier, S., Prieur, M., Dutrillaux, B., Seringe, Ph., Rossier, A. and Job, J. -C. Trisomie pour la partie distale du bras court du chromosome 3 chez trois germains. Premier exemple d'insertion chromosomique. Ann. Genet. (Paris) 15: 159-165 (1972).

73. Revazov, A. A., Derilo, T. G. and Vorsanova, S. G. On the enlarged long arm of a B-group chromosome Bq+ in a nine-year-old boy. Genetika 7(8): 170-175 (1971).

74. Rohde, R. A. and Catz, B. Maternal transmission of a new group C (6-9) chromosomal syndrome. Lancet 2: 838-840 (1964).

75. Rott, H. -D., Schwanitz, G. and Grosse, K. -P. Partielle Trisomie Cq bei balancierter B4/Cq- Translokation bei der Mutter. Z. Kinderheilkd. 109: 293-299 (1971).

76. Rott, H. -D., Schwanitz, G., Grosse, K. -P. and Alexandrow, G. C11/D13-translocation in four generations. Humangenetik 14: 300-305 (1972).

77. Sartori, A., Tenconi, R., Baccichetti, C. and Pujatti, G. Familial 4/22 translocation with partial trisomy for the short arm of chromosome 4 in two sibs. Acta Paediatr. Scand. 63: 631-635 (1974).

78. Schachenmann, G., Schmid, W., Fraccaro, M., Mannini, A., Tiepolo, L., Perona, G. P. and Sartori, E. Chromosomes in coloboma and anal atresia. Lancet 2: 290 (1965).

79. Schinzel, A. and Schmid, W. Partielle Trisomie des kurzen Arms von Chromosom 4 mit Translokation 4p-, 18q+ beim Vater. Humangenetik 15: 163-171 (1972).

80. Schinzel, A., Hayashi, K. and Schmid, W. Mosaic-trisomy and pericentric inversion of chromosome 9 in a malformed boy. Humangenetik 25: 171-177 (1974a).

81. Schinzel, A., Schmid, W. and Mürset, G. Different forms of incomplete trisomy 13 mosaicism and partial trisomy for the proximal and distal long arm. Humangenetik 22: 287-298 (1974b).

82. Schmid, W. Pericentric inversions (Report on two malformation cases suggestive of parental inversion heterozygosity). J. Genet. Hum. 16: 89-96 (1967 and 1968).

83. Schwanitz, G. and Grosse, K. P. Partial trisomy 4p with translocation 4p-, 22p+ in the father. Ann. Genet. (Paris) 16: 263-266 (1973).

84. Schwanitz, G., Schamberger, U., Rott, H. -D. and Wieczorek, V. Partial trisomy 9 in the case of familial translocation 8/9 mat. Ann. Genet. (Paris) 17: 163-166 (1974).

85. Shaw, M. W., Cohen, M. M. and Hilderbrandt, H. M. A familial 4/5 reciprocal translocation resulting in partial trisomy B. Am. J. Hum Genet. 17: 54-69 (1965).

86. Soukup, S. W., Yarema, W. and Robinow, M. A pericentric inversion of a chromosome 4 with a t(4q+;10p-) and a familial t(DqDq) in a mentally retarded girl. Humangenetik 25: 69-78 (1974).

87. Stalder, G. R., Bühler, E. M., Gadola, G., Widmer, R. and Freuler, F. A family with balanced $D^1 \rightarrow C^8$ translocation carriers and unbalanced offspring. Humangenetik 1: 197-200 (1964).

88. Steele, M. W., Pan, S., Mickell, J. and Senders, V. Trisomy for the distal half of the long arm of chromosome No. 18. J. Pediatr. 85: 827-829 (1974).

89. Šubrt, I. and Brychnáč, S. Trisomy for short arm of chromosome 20. Humangenetik 23: 219-222 (1974).

90. Surana, R. B. and Conen, P. E. Partial trisomy 4 resulting from a 4/18 reciprocal translocation. Ann. Genet. (Paris) 15: 191-194 (1972).

91. Taft, P. D., Dodge, Ph. R. and Atkins, L. Mental retardation and multiple congenital anomalies. Association with extra small metacentric chromosome. Am. J. Dis. Child. 109: 554-557 (1965).

92. Talvik, T., Mikelsaar, A. -V. N., Mikelsaar, R. V. -A., Käosaar, M. and Tüür, S. Inherited translocations in two families t(14q+;10q-) and t(13q-;21q+). Humangenetik 19: 215-226 (1973).

93. Turleau, C., de Grouchy, J., Chavin-Colin, F., Roubin, M. and Langmaid, H. Trisomie 9p: Deux nouvelles observations. Ann. Genet. (Paris) 17: 167-174 (1974).

94. Uchida, I. A. and Lin, C. C. Identification of partial 12 trisomy by quinacrine fluorescence. J. Pediatr. 82: 269-272 (1973).

95. Valdmanis, A., Pearson, G., Siegel, A. E., Hoeksema, R. H. and Mann, J. D. A pedigree of 4/18 translocation chromosomes with type and countertype partial trisomy and partial monosomy for chromosomes 18. Ann. Genet. (Paris) 10: 159-166 (1967).

96. Vogel, W., Siebers, J. -W. and Reinwein, H. Partial trisomy 7q. Ann. Genet. (Paris) 16: 277-280 (1973).

97. Wahrman, J., Atidia, J., Goiten, R. and Cohen, T. Pericentric inversions of chromosome 9 in two families. Cytogenetics 11: 132-144 (1972).

98. Walzer, S., Favara, B., Ming, P. M. L. and Gerald, P. S. A new translocation syndrome (3/B). N. Engl. J. Med. 275: 290-298 (1966).

99. Warburg, M. and Andersen, S. Ry. Ocular changes in simple trisomy and in a few cases of partial trisomy. Acta Ophthalmol. (Kbh.) 46: 372-384 (1968).

100. Wilson, M. G., Towner, J. W., Coffin, G. S. and Forsman, I. Inherited pericentric inversion of chromosome No 4. Am. J. Hum. Genet. 22: 679-690 (1970).

101. Yanagisawa, S. and Hiraoka, K. Familial C/G translocation in three relatives associated with severe mental retardation, short stature, unusual dermatoglyphics and other malformations. J. Ment. Defic. Res. 15: 136-146 (1971).

102. Yunis, J. J. and Sanchez, O. A new syndrome resulting from partial trisomy for the distal third of the long arm of chromosome 10. J. Pediatr. 84: 567-570 (1974).

103. Zaremba, J., Zdzienicka, E., Glogowska, I., Abramowicz, T. and Taracha, B. Four cases of 9p trisomy resulting from a balanced familial translocation (9;15)(q13;q11). Clinical picture and cytogenetic findings. J. Ment. Defic. Res. 18: 153-190 (1974).

CHAPTER 11

BALANCED TRANSLOCATIONS

I. OCULAR ABNORMALITIES

The ocular abnormalities associated with balanced translocations are summarized in Table 11-1. These translocations fall into two groups: (a) where parents have normal phenotype and karyotype; and (b) where one parent is phenotypically normal but carries the same balanced translocation. In the first group, the translocation is de novo and may have occurred during a very early cleavage division of the egg, or in the germ cells of one of the parents. The translocation in this case may not be balanced because the abnormal phenotype of the child leads one to suppose that there has been some loss or gain of chromosome segments. Another supposition is that there may be a submicroscopic effect with a single band containing several hundred genes and where either several genes may be lost in a translocation or the sequence of genes may have been disturbed, resulting in a phenotypic manifestation. It is surprising, in fact, that balanced translocations do not cause an apparent effect on the phenotype, although it is possible that the breakpoints occur either on genetically inactive segments or at highly repetitive sequences of DNA. A third hypothesis has been suggested, namely a position effect in that the location of one segment of a chromosome may alter the phenotype when this segment is located on another chromosome or elsewhere on the same chromosome. The position effect has not been demonstrated in man, but it has been demonstrated in the mouse, Drosophila, and in some plants.

A translocation may be apparently balanced, but sophisticated techniques may prove the existence of a deletion. Recently, Ladda and his colleagues (16) analyzed a translocation in association with aniridia and Wilm's tumour. This translocation was found to be t(8p+;11q-) by trypsin banding, but by using a computer scan they were able to detect an interstitial deletion 8p22. The second group of cases where a translocation is inherited from a phenotypically normal parent is difficult to explain satisfactorily. The abnormal phenotype in the child implies a secondary event, as for example, a paracentric inversion in one of the translocated chromosomes. There is a long account of buphthalmos associated with a chromosomal translocation (13), but the association is not significant. There is also a detailed report of albinism affecting six children in a sibship of eight (1). The mother and all her children carried the translocation, but two of her children had normal pigmentation, so that there was no association between albinism and the chromosomal abnormality.

In conclusion, it is interesting to note that mental retardation is almost the rule in these cases of balanced and unbalanced translocations. This does not mean that all subjects with mental retardation have a chromosomal abnormality, but mental retardation associated with odd facies, other congenital abnormalities, and a family history of recurrent abortions, is strongly suggestive of an associated chromosomal translocation.

TABLE 11-1

Balanced Translocations

Author	Ocular Abnormalities	Chromosome Translocation	Parental Karyotype
Lejeune et al. (17) 1968	Brushfield spots	1p+;2p-	1p+;2p- mat
Genest et al. (12) 1971	Small palpebral fissure, enophthalmos, epicanthus, hypertelorism, generalized pallor of the papilla, foveal reflex absent	2q-;Bq+	Normal
Dodson et al. (8) 1970	Small orbits, hypertelorism, antimongoloid slant, some exophthalmos	2p-;Cq+	Normal
de Grouchy and Lautmann (14) 1968	Narrow palpebral fissures, antimongoloid slant	1q-;Dq+ and 2q-;16q+	Normal
Mercer and Darakjian (19) 1962	"Eyes deep set with rather striking dark circles about them"	t2;D	Not stated
Reisman and Kasahara (23) 1968	Bilateral epicanthus, hypertelorism, alternating esotropia, superior-nasal displacement of pupils	2q-;Dq+	Normal
Goddé-Jolly et al. (13) 1970	Buphthalmos, corneal opacities, enormous papillary excavation with vessels all bunched on nasal side	1p-;Cp-q+	Normal
Dallapiccola (7) 1971	Retinitis pigmentosa, nystagmus, incomplete Laurence-Moon-Biedl syndrome	2p-;17p+	2p-;17p+ mat

TABLE 11-1 (continued)

Author	Ocular Abnormalities	Chromosome Translocation	Parental Karyotype
Ferrari and Hering (9) 1969	Hypertelorism, bilateral epicanthus, fundus appearance suggestive of blindness	1p−;17q+	Normal
Maganias et al. (18) 1967	Bilateral myopia, internal strabismus of the right eye	1?−;Gq+	Gq+mat
Rumpler et al. (26) 1967	Hypertelorism, antimongoloid slant, conjunctival hyperaemia, Brushfield spots, venous congestion	Bq−;Dq+	Bq−;Dq+ mat
Serville et al. (27) 1974	Bilateral anophthalmos, lateral facial fissure, mental retardation	t(4;14)	Normal
Beauvais et al. (3) 1969	Hypertelorism, antimongoloid slant, horizontal nystagmus, slight divergent strabismus	5q−;Dq+	5q−;Dq+ mat
Carnevale and de los Cobos (4) 1973	Hypertelorism, blue sclerae	Dq+;Dq− mosaic	Normal
Neuhäuser and Back (22) 1967	Hypotelorism, mongoloid slant, exophthalmos, conjunctivitis, bilateral corneal opacities, slightly raised intraocular pressure, blurred papillae	C−;Xp+	Normal
Bargman et al. (2) 1970	Bilateral epicanthus, narrow palpebral fissures, ?ptosis, slight hypertelorism, coarse pendular nystagmus	Cp+;Cq−	Cp+;Cq−
Hansen (15) 1970	Rotatory spontaneous nystagmus, ?blindness	Cq−;Ep+ mosaic	Normal

TABLE 11-1 (continued)

Author	Ocular Abnormalities	Chromosome Translocation	Parental Karyotype
Therkelsen et al. (28) 1971	Close-set eyes with small palpebral fissures, bilateral blepharophimosis	Presumptive C/F	Paternal C/F
André et al. (1) 1972	Sibship of 8; albinism 6/8; strabismus, 5/6; myopia, 1/6; hypermetropia, 2/6	11q-;20q+	11q-;20q+ mat
Nakagome et al. (21) 1973	Case B: antimongoloid slant	11q-;14q+	Normal
Gemme et al. (11) 1973	Slight pallor of the papilla	8q+;9q-	8q+;9q- mat
Ladda et al. (16) 1974	Bilateral aniridia, glaucoma, megacornea, anterior polar cataracts, unilateral Wilm's tumour	Apparent 8p+;11q-	
Mikkelsen et al. (20) 1973	Slightly oblique palpebral fissures, left-sided convergent strabismus with inward rotation of the eyeball	15q-;6q+	Normal
Cowie et al. (5) 1965	Mongoloid slant, pupils surrounded by light grey areola were not fixed, but very little reaction to light, nystagmus	D/F	Not studied
Wallace and Anderson (30) 1964	Eyeless orbits	(BqDq)	Not stated
Geib and Pfeiffer (10) 1969	Slight epicanthus, myopia and astigmatism	t(4p14q)	Normal
Vogel et al. (29) 1971	Hypertelorism, epicanthus	t(13qCp+)	Normal

TABLE 11-1 (continued)

Author	Ocular Abnormalities	Chromosome Translocation	Parental Karyotype
Rowley and Pergament (24) 1969	Hypertelorism, epicanthus, antimongoloid slant	(13q14q)mar+	(13q14q) mat
Rumpler et al. (25) 1969	Alternating convergent strabismus	t(14q14q)	Normal
Crandall et al. (6) 1972	Hypotelorism, enophthalmos and strabismus in two sibs	t(13q14q)	t(13q14q) mat

REFERENCES

1. André, M. J., Baron, A., Catros, A., Dejour, Kardjiev, L. and Tusques, J. Etude génétique d'une famille presentant l'association d'une translocation chromosomique et d'un albinisme universal. Bull. et Mem. Soc. Franc. Ophtal. 85: 112-119 (1972).

2. Bargman, G. J., Neu, R. L., Powers, H. O. and Gardner, L. I. A 46, XX, t(Cp+;Cq-) translocation in a girl with multiple congenital anomalies and in her phenotypically normal father 46, XY, t(Cp+;Cq-). J. Med. Genet. 7: 77-80 (1970).

3. Beauvais, P., Rumpler, Y., Ruch, J. -V., Dreyfus, J. and Haeberle, Cl. A propos d'une translocation d'une partie des bras longs d'un chromosome 5 sur les bras longs d'un chromosome du groupe D (Bq-, Dq+). Arch. Fr. Pediatr. 26: 695-704 (1969).

4. Carnevale, A. and Cobos, L. de los. A child with multiple congenital malformations and a 46, XX, t(Bq+:Dq-)/45, XX, -B, -D, +der(B), t(Bq+:Dq-) karyotype. J. Med. Genet. 10: 376-379 (1973).

5. Cowie, V., Kahn, J. and O'Reilly, J. N. Congenital abnormalities in a child with an apparently balanced karyotype carrying a reciprocal D/F translocation. Lancet 1: 1043-1044 (1965).

6. Crandall, B. F., Francke, U., Campbell, M. A. and Sparkes, R. S. Inherited t(13q14q) in two retarded sisters. Am. J. Hum. Genet. 24: 416-424 (1972).

7. Dallapiccola, B. Familial translocation t(2p-:17p+). Ann. Genet. (Paris) 14: 153-155 (1971).

8. Dodson, W. E., Kennedy, J. L., Jr. and Al-Aish, M. Acrocephalosyndactylia associated with a chromosomal translocation 46XXt(2p-:Cq+). Am. J. Dis. Child. 120: 360-362 (1970).

9. Ferrari, I. and Hering, S. E. Case report: Reciprocal translocation t(1p-:17q+) in a patient with multiple anomalies. Birth Defects: Orig. Art. Ser. 5 (pt 5): 132-133 (1969).

10. Geib, K. and Pfeiffer, R. A. Second case of chromosomal translocation t(4p14q) (In a boy with severe lyphoscoliosis and obesity of the trunk). Ann. Genet. (Paris) 12: 115-118 (1969).

11. Gemme, G., Vianello, M. G., Zera, M. and Reboa, E. Un caso di sindrome polimalformativa con traslocazione cromosomica, t(8q+;9q-). Minerva Pediatr. 25: 1338-1344 (1973).

12. Genest, P., Lachance, R., Poty, J. and Jacob, D. Autosomal translocation in a mentally retarded male child with 46, XY, t(2q-: 13q+) complement. Case report and review. J. Med. Genet. 8: 504-508 (1971).

13. Goddé-Jolly, D., Bonnenfant, F., Raoul, O., Clergue, G., Mallet, R. and Lejeune, J. A propos d'un cas de glaucome congénital associé à une anomalie chromosomique. Bull. Soc. Ophtal. Franc. 70: 875-881 (1970).

14. Grouchy, J. de and Lautmann, F. Caryotype 46, XX, 1q-, 2q-, Dq+, 16q+ chez une enfant polymalformée. Ann. Genet. (Paris) 11: 129-131 (1968).

15. Hansen, J. De novo Translokation C/E bei klinisch Atypischen e Trisomie Syndrom (Mosaik: 46, XX/46, XX, E+C-t(Cq-; Ep+). Klin. Wochenschr. 48: 1188-1192 (1970).

16. Ladda, R., Atkins, L., Littlefield, J., Neurath, P. and Marimuthu, K. M. Computer-assisted analysis of chromosomal abnormalities; Detection of a deletion in aniridia/Wilm's tumor syndrome. Science 185: 784-787 (1974).

17. Lejeune, J., Lafourcade, J., Réthoré, M. -O., Berger, R., Abonyi, D., Dutrillaux, B. and Cayroche, P. Translocation t(1p+:2p-) identique chez une femme et son fils arriérés mentaux. Ann. Genet. (Paris) 11: 117-180 (1968).

18. Maganias, N. H., Archambault, L., Becker, K. L. and Winnacker, J. L. A 1/G translocation in a member of a kindred with a marker chromosome. Arch. Intern. Med. 119: 297-301 (1967).

19. Mercer, R. D. and Darakjian, G. Apparent translocation between chromosome 2 and an acrocentric in group 13-15. Lancet 2: 784 (1962).

20. Mikkelsen, M., Dyggve, H. and Poulsen, H. (6:15) translocation with loss of chromosome material in the patient and various chromosome aberrations in family members. Humangenetik 18: 195-202 (1973).

21. Nakagome, Y., Iinuma, K. and Matsui, I. Three translocations involving C or G group chromosomes. J. Med. Genet. 10: 174-176 (1973).

22. Neuhäuser, G. and Back, F. X-Autosom Translokation bei einem Kind mit multiplen Missbildungen. Humangenetik 3: 300-311 (1967).

23. Reisman, L. E. and Kasahara, S. An unusual chromosome abnormality. 2/D translocation. Am. J. Dis. Child. 115: 625-628 (1968).

24. Rowley, J. D. and Pergament, E. Possible non-random selection of D group chromosomes involved in centric-fusion translocations. Ann. Genet. (Paris) 12: 177-183 (1969).

25. Rumpler, Y., Razanamparany, M. and Andriamiandra, A. Translocation t(DqDq) chez un jeune malgache achondroplase. Ann. Genet. (Paris) 12: 119-121 (1969).

26. Rumpler, Y., Ruch, J. V. and Beauvais, P. Translocation d'une partie des bras longs d'un chromosome 5 sur les bras longs d'un chromosome du groupe D (Bq-;Dq+) chez un enfant et sa mère. Humangenetik 4: 166-173 (1967).

27. Serville, F., Broustet, A., Peyresblanques, J. and Bouineau, J. Anophtalmie bilatérale, anomalie de la face et translocation t(4;14). J. Genet. Hum. 22: 341-351 (1974).

28. Therkelsen, A. J., Klinge, T., Henningsen, K., Mikkelsen, M. and Schmidt, G. A family with a presumptive C/F translocation in five generations. Ann. Genet. (Paris) 14: 13-21 (1971).

29. Vogel, W., Reinwein, H. and Engel, W. Fragliche Tandemtranslokation C/13 bei einem Jungen mit Missbildungen. Z. Kinderheilkd. 110: 166-174 (1971).

30. Wallace, C. and Anderson, I. F. Group B/D translocation chromosome in a case with stigmata of the D trisomy. S. Afr. Med. J. 38: 352 (1964).

CHAPTER 12

ANEUPLOID TRANSLOCATIONS

I. OCULAR ABNORMALITIES

Aneuploid translocations arise out of a 3 to 1 meiotic disjunction (see page 23) and these subjects all have either 45 or 47 chromosomes with partial trisomy, partial monosomy or tetrasomy of some chromosome regions. A recent review of the literature suggests that these translocations occur more often than was previously recognized (3).

Four reports of aneuploid translocations with associated ocular abnormalities are described here. Dinno et al. (1) described a boy with microcephaly, bilateral cleft lip and palate, hypoplasia of maxilla and mandible, hypertelorism, epicanthus, and antimongoloid slant who died at the age of nine months. The karyotype was 47,XY,t(9p+;11q+). The second report is that of Fryns et al. (2) who reported a girl aged nine with hypertelorism, the ophthalmological examination was normal. Her karyotype was 47,XX,(14q-)+ and her mother had a translocation t(14;19). A case of presumptive tetrasomy for the short arm of chromosome 9 is described by Rutten et al. (4) in a six-year-old girl who had hypertelorism, bilateral epicanthus, convergent strabismus, and high myopia. The extra chromosome present was a double 9p, which was possibly an isochromosome, but her parents had normal karyotypes. The final example of an aneuploid translocation is a partial free trisomy 13 in a girl with convergent strabismus described by Schwanitz et al. (5). Her karyotype was 47,XX,(13q-)+ and her mother had a translocation t(13;16)(q14;q24).

These four cases have been included for reference and it is very possible that there are other reports.

REFERENCES

1. Dinno, N. D., Silvey, G. L. and Weiskopf, B. 47, XY, t(9p+;11q+) in a male infant with multiple malformations. Clin. Genet. 6: 125-131 (1974).

2. Fryns, J. P., Cassiman, J. J. and Berghe, H. van den. Tertiary partial 14 trisomy 47, XX, +14q-. Humangenetik 24: 71-77 (1974).

3. Lindenbaum, R. H. and Bobrow, M. 3:1 meiotic disjunction resulting in 47- or 45-chromosome offspring. J. Med. Genet. 12: 29-43 (1975).

4. Rutten, F. J., Scheres, J. M. J. C., Hustinx, Th. W. J. and Ter Haar, B. G. A. A presumptive tetrasomy for the short arm of chromosome 9. Humangenetik 25: 163-170 (1974).

5. Schwanitz, G., Grosse, K. -P., Semmelmayer, U. and Mangold, H. Partielle freie Trisomie 13 in einer familie mit balancierter Translokation (13q-;16q+). Monatsschr. Kinderkeilkd. 122: 337-342 (1974).

EPILOGUE

It is apparent from this survey of the literature that ocular abnormalities are a common finding in deficiency and duplication syndromes, with hypertelorism, epicanthus, slanted palpebral fissures, and strabismus being most frequently reported. Other ocular abnormalities, such as ptosis, coloboma, nystagmus, and cataract, occur less often and when considered with other physical signs, these may help to distinguish the various syndromes. The accurate identification of chromosome segments is now possible with the use of the relatively new banding techniques. With further refinement of existing techniques and with the advent of new methods, it is inevitable that a host of new deficiency and duplication syndromes will be detected. Although the references in this review are complete only to the end of 1974, it is already apparent that new syndromes are being described. It is reasonable to suppose that in some cases, the new syndromes will differ from those already reported only in small details in the specific phenotype.

The comprehensive listing of all phenotypic abnormalities is of great value, and in this respect the ophthalmologists have much to offer the geneticists. It is only by comprehensive and accurate listing of congenital abnormalities that some advance may ultimately be made towards gene mapping. For example, ophthalmologists have already made a significant contribution to the assignment of the locus of a gene for normal retinal development on the long arm of chromosome 13, and to other gene loci on the X chromosome.

It would be foolish to assume, however, the existence of a single gene responsible for epicanthus or strabismus, to cite but two conditions which appear to be polygenic in origin. The study of deficiency and duplication syndromes is concerned primarily with groups rather than with individual genes, and relatively little is known of the precise action of a group of genes beyond its phenotypic effect. It may be possible, in time, to determine gene dosage effects, and the comparison of the trisomic and deficiency states will be most valuable, and these comparisons may also be fruitful in demonstrating position effects. The field of research for geneticists is vast, and it is to be hoped that ophthalmologists can be persuaded that they have a considerable and meaningful contribution to make in this fascinating topic.

GLOSSARY

Acrocentric: descriptive of a chromosome with its centromere towards one end of its length.

Anaphase: one of the stages of cell division.

Aneuploidy: any deviation from the normal number of chromosomes.

Aniridia: a rudimentary fringe of iris, appearing as though the iris is absent.

Antimongoloid slant: a downward and outward slant of the palpebral fissures.

Ascertainment: the manner in which a patient is first discovered.

Autoradiography: a method of identifying chromosomes by labelling with a radioactive isotope.

Autosome: one of the somatic chromosomes (as opposed to one of the two sex chromosomes).

Anisocoria: pupils of unequal size.

Barr body: the deeply staining (heterochromatic) X chromosome present in the interphase nuclei of normal females.

Blepharochalasis: a condition characterized by an atrophy and relaxation of the tissues of the upper lids, often appearing like a ptosis.

Brushfield spots: golden or white speckling of the iris.

Cataract: an opacity of the lens.

Centromere: the primary constriction of a chromosome.

Chiasma: a point at which chromosomes exchange genetic material during meiosis.

Chromatid: one of the two identical strands which constitute a chromosome during cell division.

Clinodactyly: bent fingers.

Clone: one of a group of genetically identical cells derived from a single cell.

Coloboma: a defect due to failure of closure of the fetal cleft in the lower part of the eye.

Congenital: present at birth.

Crossing-over: the process whereby chromosomes of maternal and paternal origin exchange genetic material at meiosis.

Deficiency: the absence of a chromosome segment, this segment may be lost (see deletion), or present on another chromosome as in a translocation, or lost as in ring chromosomes.

Deletion: the loss of a chromosome segment.

Dermatoglyphs: the genetically determined dermal patterns of palms and soles.

Deuteranopia: an anomaly of colour vision where there is no sensation of green.

Diploid: the normal (double) number of chromosomes present in a somatic cell.

Dominant: describes a trait or condition for which the gene responsible has to be present in a single dose.

Duplication: the presence of a segment of a chromosome in excess.

Ectropion: the turning outward (eversion) of the lid margins.

Enophthalmos: abnormally deep-set eyes.

Epicanthus: a half-moon shaped fold of skin running downwards at the side of the nose.

Exophthalmos: abnormal protrusion of the eyeball.

Fovea: a shallow depression at the centre of the macula at the level of the lower margin of the optic disc, where the visual cells are all cones and where vision is most acute.

Gamete: an ovum or a sperm.

Genotype: the genetic constitution of an individual.

Glaucoma: pathologically raised intraocular pressure, associated with damage to vision.

Haploid: the normal (single) number of chromosomes present in a germ cell (sperm or ovum).

Heterochromatin: chromatin which, when nuclear stains are used, stains more intensely and is visible during interphase.

Heterozygote: an individual in whom the gene responsible for a certain condition is acting in a single dose.

Homozygote: an individual in whom the gene responsible for a certain condition is acting in double dose.

Hypertelorism: increased distance between the orbits.

Hypotelorism: decreased distance between the orbits.

Interphase: one of the stages during which the cell is at rest and not dividing.

Inversion: the rotation of an interstitial fragment of a chromosome followed by reunion of the broken ends.

Isochromosome: a chromosome formed of identical long or short arms after a transverse break of the centromere.

Karyotype: the chromosome complement of an individual.

Keratoconus: a condition where the central part of the cornea is distorted into the shape of a cone due to an increased curvature of the cornea.

Lid lag (von Graefe's sign): the upper lid lags behind the globe when the patient looks down (commonly a sign of hyperthyroidism.)

Locus: the site on a chromosome occupied by a gene.

Macula: an area of retina about 3 mm in diameter surrounding the fovea and situated temporal to the optic disc.

Megalocornea: an enlarged circumference of the cornea.

Meiosis: a form of cell division by reduction occurring during the formation of primary germ cells whereby the chromosome number is halved.

Metacentric: descriptive of a chromosome with the centromere near the middle of its length.

Metaphase: one of the stages of cell division.

Microcornea: an abnormally small cornea.

Microphthalmos: an abnormally small eye.

Mitosis: the process of cell division of all somatic cells whereby each daughter cell retains the same number of chromosomes as the parent cell.

Mongoloid slant: an upward and outward slant of the palpebral fissures.

Mosaic: the state where two or more cell lines with different chromosome complement are present in an individual or organ. The term mixoploidy is also used to describe this state.

Nondisjunction: the failure of two homologous chromosomes to pass to opposite poles of the cell at meiosis, or the failure of two chromatids to pass to opposite poles of the cell at mitosis or at the second division of meiosis.

Nystagmus: fine rhythmic oscillations of the eyes which may be pendular or jerky.

Optic atrophy: a condition where the nerve fibres have atrophied.

Optic disc: the head of the optic nerve which is visible ophthalmoscopically.

Papilla: see optic disc.

Phenotype: the physical constitution of an individual.

Proband: the first affected member of a family who is examined.

Protanopia: an anomaly of colour vision where there is no sensation of red.

Ptosis (blepharoptosis): drooping of the upper lid.

Pupillary membrane: an embryonic structure which originally forms the anterior layer of the iris but which normally atrophies before birth.

Recessive: describes a trait or condition for which the gene responsible has to be present in a double dose.

Replication: the mechanism by which the DNA molecule divides and new DNA is synthesized.

Retinal dysplasia: a pathological condition characterized by rosette formation resulting from disturbed retinal growth and differentiation.

Retinoblastoma: a malignant neoplasm affecting the retina, occurring congenitally or in very young patients.

Sclerocornea: an opacification of the peripheral parts of the cornea so that it appears to merge with the sclera.

Strabismus: an abnormality of the eyes in which the visual axes do not meet at the desired objective point.

Submetacentric: descriptive of a chromosome with its centromere to one side of the centre.

Telomere: a structure at the ends of the arms of a chromosome.

Translocation: the transfer of a segment or a whole chromosome onto another chromosome.

Trisomy: the presence of an extra chromosome of any one pair.

Tunica vasculosa lentis: an embryonic vascular tunic surrounding the lens.

Uvea: the vascular coat of the eye comprising the choroid, ciliary body and the iris.

Zygote: the fertilised ovum.

AUTHOR INDEX

Numbers in parentheses are reference numbers and indicate that an author's work is referred to although his name is not cited in the text. Underlined numbers give the page on which the complete reference is listed.

A

Aarskog, D., 111(1), 115, 127, 178 192
Aase, J. M., 41(34), 65
Abbassi, V., 162(55), 167
Abbott, J., 31(3), 32
Abe, T., 183(28), 194
Abonyi, D., 78(28), 81(28), 97, 172(71), 183(71), 198, 203(17), 208
Abrams, C. A. L., 78(2), 81(2), 91(2), 95
Abramowicz, T., 184(103), 200
Abreu, M. C., 49(17), 64
Acoto, T., 162(55), 167
Adam, M. S., 78(1, 22), 80, 81(22), 95, 96
Adams, F. G., 36(115), 41(115), 72
Ahmed, F., 137, 137
Aichmair, H., 101(30), 103(30), 104(30), 129
Al-Aish, M. S., 139(1, 2), 141, 142, 145, 203(8), 207
Albert, A., 152(2), 163
Albert, D. M., 29(4), 32, 86(20), 93(20), 96
Albores, J. M., 108(14), 128, 174 (15), 193
Alexander, D. S., 153(19), 164
Alexandrow, G., 170(76), 185(76), 198
Alfi, O. S., 73, 75, 75, 182, 192
Allderdice, P. W., 40(114), 41(68), 44(8), 58(8, 114), 59(8), 63, 68, 71, 78(2), 81, 91, 95, 116(20), 128
Allen, F. H., Jr., 41(68), 68, 78(2), 81(2), 91(2), 95, 141 (20), 146
Al-Salihi, F. L., 116(20), 128
Altrogge, H. C., 36(1), 41(81), 59, 60, 63, 69
Ananjev, E. V., 190(57), 196
Anders, G. J. P. A., 118(40), 130, 179(34), 194
Andersen, S. Ry., 182, 200
Anderson, I. F., 205, 209
Anderson, R., 115(29), 123(29), 129
Anderson, V. E., 103(39), 104(39), 107(39), 130
André, M. J., 202(1), 205, 207
Andriamiandra, A., 206(25), 209
Ansehn, S., 50(6), 63
Antich, J., 47(2, 93), 56, 63, 70
Arakaki, D. T., 103(2), 127
Arcadi, J., 139(26), 141(26), 146

Archambault, L., 139(26), 141(26), 146, 204(18), 208
Arias, D., 40, 63
Armendares, S., 83(46), 98, 140 (28), 143, 143(28), 145, 147, 181(12), 192
Arroyo, H., 158(1), 161(40), 163, 166
Arthuis, M., 34(33), 36(33), 46(33), 49(33), 65, 101(44, 101), 103 (101), 105(44, 101), 130, 135
Asper, A. C., 41(34), 65
Atidia, 191(97), 200
Atkins, L., 73(10), 75(10), 76, 86, 86(25), 96, 98, 115(29), 123(29), 129, 170(7), 174(91), 178(7), 192, 199, 201(16), 205(16), 208
Aubert, L., 158, 161(40), 163, 166
Auger, C., 121(34), 129
Aula, P., 103(62), 109(62), 121, 127, 132
Aussannaire, M., 78(44, 48), 82(44), 83(48), 95
Ayraud, N., 81, 95, 107, 108(104), 127, 135

B

Bach, Ch., 56, 63, 190(8), 192
Baccichetti, C., 170(77), 180(77), 183, 192, 198
Bachelet, Y., 162(27), 165
Back, F., 157, 166, 204, 208
Baheux, G., 121(66), 132
Baheux-Morlier, G., 62(109), 71
Bain, A. D., 80, 95, 182(35), 194
Baitsch, H., 33(118, 120), 34(120), 36(118, 120), 38(118, 120), 51 (7), 63, 72, 141(46), 148
Baker, E., 36(11), 40(11), 63
Balci, S., 142(35), 147, 155(54), 167
Barber, L., 143(37), 146
Barcinski, M. A., 49(17), 64
Bargman, G. J., 36(53), 39(53), 67, 204, 207
Baron, A., 202(1), 205(1), 207
Barr, M., 59, 66
Barr, M. L., 156(7), 163
Barrett, R. V., 123(25), 129
Barton, M. E., 107(69), 132
Bartsocas, C., 115(29), 123(29), 129
Bass, H. N., 140, 145
Bass, L. W., 47(112), 55(112), 71
Batts, J. A. Jr., 103(5), 127
Bauchinger, M., 141, 145
Bauer, H., 56(99), 70
Bearn, A. G., 141, 145
Beaudoing, A., 36(45), 39(45), 66
Beauvais, P., 204(26), 204, 207, 209
Bec, P., 154(6), 163
Becker, K. L., 152, 163, 204(18), 208
Belotti, B. M., 55(9), 63
Bender, K., 178, 192
Benirschke, K., 74(15), 76, 141 (20), 146
Benson, P. F., 153, 163
Beráńková, J., 109, 134
Berg, J. M., 50, 63
Berger, R., 36(24), 42(24), 44(59), 46(24, 60), 47(24, 58), 49(59), 50(56, 60), 51(57), 64, 67, 78(28), 79(11), 81(28), 85(11), 88(11), 95, 97, 106(64), 111 (32, 63), 114(63), 116(65), 121 (66), 129, 132, 139(26), 141 (26), 146, 161(47), 166, 172 (71), 183(71), 185(51), 196, 198, 203(17), 208
Berghe, H. van den, 43(25), 65, 84(12), 95, 155(8), 163, 170,

(Berghe, H. van den)
178, 183(26), 192, 194, 211(2), 212
Berlin, C. M., 83(42), 98, 125(89), 134, 143(32), 147
Berlow, S., 119(85), 124(85), 134
Bergman, S., 50, 63
Bernard, P. -J., 56(106), 71
Bernard, R., 121, 127, 159(26), 165
Bettecken, F., 51, 63
Beun, J., 108(70), 132
Beyer, P., 174(6), 192
Bias, W. B., 115(8), 127
Bier, L., 179(56), 196
Bilchick, R. C., 77(5), 78(5), 79, 83, 95
Biles, A. R. Jr., 82, 95
Blank, C. E., 142, 145, 180(63), 197
Blěhová, B., 36(108), 40(108), 71
Bloom, G. E., 78, 83(18), 95, 96
Blumina, M. G., 107, 127, 161(42), 166, 190, 197
Bobrow, M., 143(27), 146, 211(3), 212
Boeswillwald, M., 44(59), 49(59), 67, 172(71), 183(71), 198
Bohe, B., 78(15), 83(15), 96
Böhm, R., 142, 145
Boisse, J., 106(64), 132
Bonnenfant, F., 202(13), 203(13), 208
Bonnet, H., 142(11), 145
Bonnet, P., 155(33), 165
Bonnette, J., 73(8), 74(8), 76, 106 (42), 130
Böök, J. A., 170(7), 178, 192
Borgaonkar, D. S., 115(8), 127
Borkowf, S. P., 115, 127
Boschann, H. W., 156(7), 163
Bouchard, R., 137(3), 138
Boué, J., 29(4), 32, 190, 192
Bove, K. E., 155, 163
Bouineau, J., 204(27), 209
Boutu, F., 60(18), 64
Bovier-Lapierre, M., 185(50), 196
Bowen, P., 153(19), 164
Boyd, E., 158(20), 164
Brabosa, L. T., 49(17), 64
Brandt, N. J., 59(86), 69
Breg, W. R., 39(67), 41(68), 44, 46(29), 47(107), 53(69), 55(107), 58, 59, 63, 71, 86(20), 93(20), 96, 139(17), 144(17), 146
Briggs, D. K., 156, 163
Briggs, J. H., 152(24), 165
Brihaye, M., 123, 127
Brostrøm, K., 144(21), 146
Broustet, A., 204(27), 209
Broyer, M., 139(26), 141(26), 146
Brusquet, Y., 56(106), 71, 121 (6), 127, 159(26), 165
Brychňač, S., 187, 199
Buchinger, G., 62(51), 66
Buentello, L., 83(46), 98, 143(3), 145, 181(12), 192
Buffoni, L., 55, 63
Bühler, E., 175(42), 186(42), 195
Bühler, E. M., 105, 127, 142(47), 148, 174(9), 175(9), 186(87), 192, 199
Bühler, U. K., 105(10), 127, 142 (47), 148
Bulteel, M. F., 123(23), 128, 161 (15), 164, 181(19), 193
Bunting, K. W., 161(18), 164
Büsse, M., 62(51), 66
Butler, L. J., 36(62), 42, 67, 73(3), 74(3), 75, 183, 190, 192
Büttinghaus, K., 186(64), 197

C

Cabrol, C., 57(79), 69, 78(15), 83(15), 96, 109(35), 129

(Cabrol, C.)
83(15), 96, 109(35), 129
Cagianut, B., 85(8), 88, 92, 95
Cahalane, S. F., 81(30), 97
Calmettes, L., 154, 163
Cameron, A. H., 107(69), 132
Campbell, M. A., 206(6), 207
Campbell, W. A. B., 143(31), 147
Canet, J., 184, 195
Cant, J. S., 47(64), 55(64), 68
Cantu, J. -M., 181, 192
Cantu-Garza, J. -M., 143(3), 145
Cao, A., 180(29), 194
Capdeville, C., 107(4), 127
Capoa, A. de, 39(67), 40(114), 41 (68), 44(8), 54(8, 114), 58(8), 59(8), 63, 68, 71
Capra, L., 180(18), 193
Capraro, V. J., 103(18), 120(18), 125(18), 128, 162(55), 167
Carlon, N., 158(1), 161(40), 163, 166
Carnahan, L. G., 123(25), 129
Carnavale, A., 204, 207
Carpenter, G. G., 50(88), 69
Carpentier, S., 172(71), 178(72), 183(71), 198
Carr, D. H., 77(56), 99, 119(107), 135, 156, 163
Carrel, R. E., 78(49), 80(49), 98, 187, 193
Carter, C., 61(101), 70
Carter, R., 36(11), 40, 63
Casey, M. D., 181(54), 196
Cassiman, J. J., 84(12), 95, 183 (26), 194, 211(2), 212
Casteels-Vandaele, M., 155, 163
Cathelineau, L., 62(109), 71
Catti, A., 47(12), 60, 64
Catros, A., 202(1), 205(1), 207
Catz, B., 182, 198
Cavazzuti, G., 178(25), 194
Cayroche, P., 203(17), 208
Ceccarelli, M., 51(28), 65
Cedrato, A., 108(14), 128
Cenani, A., 115, 123, 127
Chaban, J., 155, 167
Chabrune, J., 121(66), 132
Challacombe, D. N., 36(111), 41(111), 71, 140(8), 142, 145
Chang, P., 124, 128
Chapelle, A. de la, 143(24), 146, 152, 163, 180, 193
Charles, J. -M., 109(35), 129
Chauvel, P. J., 143, 145
Chavin-Colin, F., 184(93), 199
Ch'Eng, L. Y., 60, 64
Chen, A. T. L., 31(10), 32
Chieri, P. R. de, 108, 128, 174 (15), 193
Chinitz, J., 58(100), 70
Chitham, R. G., 54, 68
Chown, B., 111(81), 115(81), 133
Christensen, K. R., 123, 128
Christodorescu, D., 158(50), 167
Chung, C. Y., 139(34), 140(34), 141(34), 147
Chutorian, A. M., 39(67), 68
Ciovirnache, M., 158(50), 167
Citoler, P., 41, 64
Clarke, G., 178, 193
Clergue, G., 202(13), 203(13), 208
Cobos, L. de los, 204, 207
Coetsier, H., 154(30), 165
Coffin, G. S., 78(9), 82, 95, 179(100), 200
Cohen, M. M., 103(16, 18), 109, 120(18), 125, 128, 140(10), 141, 145, 153(29), 161, 162 (55), 163, 165, 167, 170(85), 179(85), 190, 193, 199
Cohen, M. M. Jr., 61(101), 70
Cohen, T., 191(97), 200
Cohen-Solal, J., 122(43), 130
Coles, J. W., 103(5), 127
Colover, J., 46, 61, 64

Colver, D., 79(16), 83(16), 86(16), 96
Comley, J. A., 46(15), 61(15), 64
Condron, C. J., 73(4), 75(4), 75
Conen, P. E., 180, 199
Constantinescu, E., 158(50), 167
Cooke, P., 36(115), 41(115), 57, 65, 72
Cooper, H. L., 33, 33(39), 34(38), 65, 66
Corcoran, P. A., 78(13), 80(13), 95
Cordier, J., 154(57), 157, 163, 164, 167
Corone, P., 190(39), 195
Cortesi, M., 124, 128
Cossu, P., 180(29), 194
Cotton, J. E., 108(49), 131
Court-Brown, W. M., 153, 157, 164
Cousin, J., 60(18), 64
Couturier, J., 183(69), 197
Covic, M., 180(18), 193
Cowie, V., 205, 207
Cramer, H., 84(19), 96
Crandall, B. F., 73(1), 75(1), 75, 103(67), 108(67), 132, 140(4), 145, 206, 207
Creery, R. D. G., 181(36), 195
Croquette-Bulteel, M. F., 123(22), 128
Croudace, J., 181(36), 195
Cruveiller, J., 78(28), 81(28), 97, 139(26), 141(26), 146, 172(71), 183(71), 198
Cruz, F. de la, 139(2), 141(2), 145
Curran, J. P., 116, 128

D

Dalaut, J. J., 157(49), 167
Dallaire, L., 36(16), 40, 64
Dallapiccola, B., 34(92), 39(92), 52(91), 70, 170(67), 180, 188(67), 193, 197, 203, 207
Darakjian, G., 203, 208
Darcourt, G., 107(4), 127
Darnell, A., 139(34), 140(34), 141(33, 34), 147
Dartnall, J. A., 181(36), 182 (37), 195
Davenport, S. L. H., 83(42), 98, 125(89), 134, 143(32), 147
David, M., 85(26), 93(26), 96
Davidson, R. G., 190, 193
Davidson, W. M., 111(21), 115 (21), 128
Davies, J., 123(106), 135
Davies, P., 178(16), 180(22), 193
Davis, J., 41(68), 68
Davis, J. G., 78(2), 81(2), 91(2), 95
Day, E. J., 111(21), 115, 128
Day, R. W., 86(10), 95
deAlmeida, J. C. C., 49, 64, 152(24), 165
Debauchez, C., 121(45), 130
Debeugny, P., 123(22), 128
Decourt, J., 151, 152, 164
Dedrato, A., 108(14), 128
Dejour, Mme., 202(1), 205(1), 207
Delaney, M. J., 36(78), 39(78), 69
Delbeke, M. J., 86(40), 93(40), 98
Delhanty, J. D. A., 50(5), 63, 170(52), 190(52), 196
Delmas-Marsalet, Y., 123(22, 23), 128
Deluchat, Ch., 29(4), 32
Deminatti, M., 60, 64, 123, 128, 161, 164, 181, 193
Dent, T., 170(52), 190(52), 196
Deodati, F., 154(6), 163

Derencsenyi, A., 73(1), 75(1), 75, 92(41), 98
Derilo, T. G., 181(73), 198
Deroover, J., 84(12), 95
Despotovic, M., 78(33), 82(33), 97
Destiné, M. L., 111(24), 115, 128
Dharmkrong-At, A., 83(18), 96
Dhermy, P., 78(48), 83(48), 98
D'Hont, G., 122(51), 131
Dietze, M. R., 153(29), 165
DiGeorge, A. M., 50(88), 69, 111 (24), 115(24), 128
Dignan, P., 174(31), 194
Dinno, N. D., 211, 212
Dodge, Ph. R., 174(91), 199
Dodson, W. E., 203, 207
D'Oelsnitz, M., 107(4), 127
Doleschel, W., 101(30), 103(30), 104(30), 108(30), 119(30), 129
Donnell, G. N., 73(1), 75(1), 75, 182(2), 192
Dossa, D., 142(11), 145
Drennan, J., 79(16), 83(16), 86(16), 96
Dreyfuss, J., 204(3), 207
Dubois, C., 56(113), 71
Dubois, R. S., 124(72), 132
Ducas, J., 78(34), 80(34), 97
Ducasse, G. C., 31(3), 32
Duke-Elder, W. S., 37(19), 64
Dumar, K. W. Jr., 49, 64, 123, 129
Dupont, A., 174(47), 186(47), 196
Dupuis, C., 123(23), 128, 181(19), 193
Dutrillaux, B., 78(28), 81(28), 86 (45), 97, 98, 170(32, 67), 172 (71), 173(70), 178(25, 72), 179 (32), 180(18, 68), 183(69, 71), 184(70), 185(50), 187, 188(67), 193, 194, 196, 197, 198, 203 (17), 208
Dyggve, H. V., 34(21), 38, 64, 205(20), 208
Dyke, H. E. van, 105, 129

E

Eades, S. M., 183(10), 192
Ebbin, A. J., 103(33), 129, 186(27), 194
Edwards, J. H., 107(69), 132, 180, 193
Edwards, R. H., 92(37), 97
Eggermont, E., 43(25), 65
Ehrhardt, A. A., 162(55), 167
Eidelman, A. I., 47(107), 55(107), 71
Elfving, J., 139(15), 143(15), 146
Ellebjerg, J., 156(44), 166
Eller, E., 159, 164
Elliott, D., 73(4), 75, 75
Ellis, E., 117(97), 134
Ellis, E. F., 124(72), 132
Ellis, J. R., 157(39), 166
Ellis, P. M., 47(64), 55(64), 68
Emberger, J. -M., 142, 145
Emerit, I., 190(39), 195
Endo, A., 142, 145
Engel, E., 46, 53, 55(103), 64, 71, 73(14), 74(14), 76, 111 (77, 81), 115(81), 117, 123 (106), 123, 133, 135, 153, 157, 161, 164
Engel, W., 173(43), 183(43), 195, 205(29), 209
Engle, M. E., 40(3), 63
Eriksson, B., 181, 193
Eygen, M. van, 170(5), 178(5), 192

F

Faed, M. J. W., 110(27), 118, 129, 137, 138
Faint, S., 103(28), 105, 129
Fareed, N., 122(84), 133
Farnir, A., 123(9), 127
Farouz, S., 155(33), 165

Faunch, J. A., 50(5), 63
Favara, B., 178(98), 200
Feingold, M., 115, 123, 129
Feldman, G. V., 36(78), 39(78), 69
Ferguson-Smith, M. A., 47(64), 55(64), 68, 153, 158, 164
Ferguson-Smith, M. E., 158(20), 164
Ferrand, J., 173(70), 184(70), 197
Ferrari, I., 204, 207
Ferrier, P., 153(51), 156, 164, 167
Ferrier, S., 153(51), 167
Finley, S. C., 170(40), 179(40), 195
Finley, W. H., 170(40), 179(40), 195
Firschein, I. L., 33(39), 38(39), 46(69), 53(69), 66, 68
Fischer, M., 154, 164
Fischer, P., 101, 103(30), 104, 108, 119, 129
Fisher, G. W., 161, 165
Fitzmaurice, F., 108(49), 131
Flatz, S. D., 181(24), 194
Flodstrom, I., 50(6), 63
Folger, G. M., 83(18), 96
Fonatsch, C., 181, 194
Forabosco, A., 170(32, 67), 178, 179(32), 188(67), 194, 197
Forbes, A. P., 150, 153, 157, 164, 166
Ford, C. E., 152, 165
Ford, E. H. R., 34, 64
Forsman, I., 179(100), 200
Fossati, P., 161(15), 164
Fournier, P., 122(56), 131
Fraccaro, M., 117, 129, 156(38), 161(37), 166, 174(78), 180(22), 181(23), 182(53), 193, 196, 198
Fraisse, J., 73, 74, 76
France, N. E., 73(3), 74(3), 75, 183(10), 192
Franchimont, P., 158(1), 163
Francke, U., 206(6), 207
François, J., 36, 42, 46, 47(24), 64, 78, 85(11), 88(11), 95, 111, 129
Frankenburg, W., 159(16), 164
Franceschini, F., 34(92), 39(92), 52(91), 70
Franklin, M., 140(23), 143(23), 146
Fredga, K., 139(16), 141(16), 146
Freuler, F., 186(87), 199
Freycon, F., 73(5), 74(5), 76
Friedrich, F., 101(30), 103(30), 104(30), 108(30), 119(30), 129
Friedrich, U., 31, 32, 75, 76, 123(15), 127, 144(21), 146
Friis-Hansen, B., 59(86), 69
Frøland, A., 156(44), 166, 174 (47), 186(47), 196
Fryns, J. P., 43, 65, 84, 95, 170(5), 178(5), 183, 192, 194, 211, 212
Fuhrmann, W., 142, 145
Fujimoto, A., 85(59), 86(59), 93(59), 99, 103(33), 129, 186, 194
Fujita, H., 183, 194
Furbetta, M., 180, 194
Furuyama, 53(47), 66, 183(28), 194

G

Gabilan, J. C., 51(32), 65
Gacs, G., 73, 74, 76
Gadola, G., 186(87), 199
Galvis, A. E., 124(82), 133
Gardner, L. I., 36(53, 54), 39 (53, 54), 55(96), 66, 70, 124(82), 133, 152(59), 155, 157(35), 165, 167, 204(2), 207

Garson, O. M., 43(46), 66
Gartler, S. M., 156(21), 164
Garzicic, B., 78(33), 82(33), 97
Gaskill, C., 49(20), 64
Gauld, I. K., 80(4), 95
Gaulme, J., 103(71), 109(71), 132
Gautier, M., 46(60), 50(56, 60), 56(4), 63, 67, 190(8), 192
Geib, K., 205, 207
Gemme, G., 205, 207
Genest, P., 51, 65, 121, 129, 203, 208
Gerald, P. S., 30, 31, 32, 78(7), 78, 80, 95, 111, 111(109), 114 (109), 117, 135, 178(98), 200
Gerlinger, P., 54(98), 70
German, J. L., 40(3), 49, 63, 65, 141, 145
Gersht, N., 103(48), 130
Gey, W., 55(119), 72, 92, 96
Giblett, E. R., 141(19), 142(45), 146, 148
Gilbert, Y., 36(45), 39(45), 66
Gilgenkrantz, S., 54(80), 57(79), 69, 78(15, 34), 80(34), 83, 96, 97, 107, 109, 129, 130, 157 (11, 49), 163, 167
Gindilis, V. M., 190(57), 196
Ginsberg, J., 36(29), 43, 65, 174 (31), 174, 194
Giorgi, P. L., 51, 65
Giovanelli, G., 170(32, 67), 179, 180, 188(67), 194, 197
Girard, J., 174(6), 192
Giraud, F., 56(106), 71, 121(6), 127, 159, 162, 165
Gleissner, M., 110, 117, 130
Glogowska, I., 184(103), 200
Gloor, R., 56(113), 71
Go, S., 117(97), 124(72), 132, 134
Goddé-Jolly, D., 202(13), 203, 208
Godeneche, P., 103(71), 108(70), 109(71), 132
Goedde, H. W., 36(1), 59(1), 60(1), 63
Goffaux, P., 84(12), 95
Goiten, R., 191(97), 200
Goldberg, B., 77(56), 99, 119 (107), 135
Goldsmith, L. A., 139(2), 141(2), 145
Goll, U., 81(54), 83(53), 99
Golob, E., 101(30), 103(30, 38), 104(30), 108(30), 119(30), 129, 130
Gonzaga, M., 49(17), 64
Goodman, R. M., 153(19), 164
Gordon, R. R., 36(115), 41(115), 52, 57, 65, 67, 72, 79(16), 83(16), 86(16), 96
Gorlin, R. J., 103, 104, 107, 130
Gorman, L. Z., 111(110), 115 (110), 135, 178(4), 192
Goroshenko, Ju. L., 93(39), 97
Gosselin, B., 181(19), 193
Gough, M. H., 153(3), 163
Gouw, W. L., 118, 130, 179, 194
Grace, E., 79, 83, 86, 96, 182, 194
Gray, B., 158(20), 164
Gray, J. E., 181, 182, 195
Gray, M., 34(102), 38(102), 70
Green, W. R., 103(5), 127
Greenwood, R. D., 140(14), 143, 146
Grinberg, K. N., 159(43), 166
Gripenberg, L., 139(15), 143(15), 146
Gripenberg, U., 121(3), 127, 139 (15), 143, 146
Groot, C. J. de, 179(34), 194
Gropp, A., 41(14), 64
Grosse, K. -P., 170(76), 182(75), 185(76), 198, 199, 211(5), 212
Grotsky, H., 46, 60, 65
Grouchy, J. de, 34(33), 36(33), 46 (60), 49(27), 49, 50(60), 51, 65, 67, 73(8), 74(8), 76, 91,

(Grouchy, J. de)
96, 101, 101(101), 103(101), 105, 105(101), 106, 110, 111, 114, 117(55), 120, 121, 122, 130, 131, 135, 137, 138, 184 (93), 184, 190, 195, 199, 203, 208
Grumbach, M. M., 156(7), 163
Guerrier, G., 73(11), 74(11), 76
Gulotta, F., 41(14), 64
Gumminger, G., 107(96), 134
Gustavson, K. -H., 170(40), 179, 195
Guthrie, R. D., 41, 65

H

Haeberle, Cl., 204(3), 207
Haentjens, P., 122, 122(51), 130, 131
Halbrecht, I., 103(48), 130
Hall, B., 139(16, 34), 140(34), 141 (33, 34), 141, 147
Hall, M. E., 190(11), 192
Halldórsson, S., 189(41), 191(41), 195
Halloran, K. H., 139(17), 144, 146
Hamerton, J. L., 31, 32, 34, 65, 140, 146, 149(28), 150(28), 156(38), 165, 166
Hansen, J., 204, 208
Hansson, K. M., 78(51), 81(51), 98
Harlan, W. L., 108, 131
Harnden, D. G., 153(13), 157(13), 164
Harris, F., 180(63), 197
Hart, M. G., 78(22), 81(22), 96
Hart, Z. H., 153, 165
Hartley, B. M. E., 78(15), 83(15), 96
Hartman, 73(11), 74(11), 76
Hartung, M., 56(106), 71, 121(6), 127, 159(26), 162(27), 165
Haslund, J., 154(22), 164
Hastings, C. P., 46(22), 53(22), 64
Hatashita, A., 92(41), 98
Haukskóttir, H., 189(41), 191, 195
Hautschteck, E., 175, 186, 195
Hayashi, K., 118(95), 134, 191 (80), 198
Hayez-Delatte, F., 55(50), 66
Hayles, A. B., 152(2), 163
Hayman, D., 36(11), 40(11), 63
Hazard, G. W., 74(2), 75
Heath, C. W. jun., 83(18), 96
Hecht, F., 36(36), 40, 65, 140 (28), 141, 142(45), 143(28), 147, 148
Heller, R. H., 74(13), 76, 153 (19), 164
Hellmich, E., 47(95), 58(95), 70
Hempel, J. M., 50(65), 68
Hendryckx, J., 54(80), 69
Henningsen, K., 205(28), 209
Hering, S. E., 204, 207
Herrault, A., 122(43), 130
Higdon, S. H., 111(81), 115(81), 133
Hijmans, J. C., 51, 65
Hildebrandt, H. M., 170(85), 179 (85), 199
Hiraoka, K., 181, 200
Hirsch, W., 144(37), 147
Hirschhorn, K., 33, 34(38), 38, 46(31, 69), 53(69), 60(31), 65, 66, 68, 186(45), 190(8), 192, 195
Hirth, L., 36(1), 59(1), 60(1), 63
Hjelt, L., 121(3), 127
Hoefnagel, D., 74(15), 76, 141, 146
Hoehn, H., 117, 131, 173(43), 178, 183, 183(69), 195, 197
Hoeksema, R. A., 115(105), 135, 187(95), 200
Holbek, S., 144, 146

Holboth, N., 34(40), 51, 66, 185, 195
Hollowell, J. G., 83, 96
Holt, S. B., 178(16), 193
Homma, T., 53(47), 66
Hoo, J. J., 84, 96, 187, 196
Hooft, C., 122, 131, 154, 165
Howard, R. O., 29, 32, 44(41), 45, 46, 61, 62, 66, 86(20), 93, 96
Howlett, R. M., 36(111), 41(111), 71
Hsu, L. Y. F., 46(31), 60(31), 65, 186, 195
Hulten, M., 117(31), 129, 181(33), 193
Humbert, J. R., 124(72), 132
Hürter, P., 181(24), 194
Hustinx, T. W. J., 51, 66, 185, 195, 211(4), 212
Hyman, G. A., 86, 99

I

Iinuma, K., 205(21), 208
Ikkos, D., 161(37), 166
Inhorn, S. L., 92(37), 97
Insley, J., 107(69), 111(52), 115, 131, 132
Ivemark, B. I., 117(31), 129

J

Jackson, L., 59, 66
Jacob, D., 203(12), 208
Jacobs, P. A., 31, 32, 153(13), 157(13), 164
Jacobsen, P., 103(53), 107, 117, 123(15), 128, 131, 174, 185(44), 186, 195
Jacobson, C. B., 83(42), 98, 125 (89), 134, 143(32), 147
Jacoby, N. M., 73(3), 74(3), 75
Jacqucloot, N., 60(18), 64
Jagiello, G. M., 157, 165
Jalbert, P., 36(45), 39, 66
Jalling, B., 170(40), 179(40), 195
Jancar, J., 159, 165
Jarrett, T. E., 47(82), 54(82), 69
Jean, R., 142(11), 145
Jenkyn, J., 182(60), 197
Jensen, R. D., 86, 96
Jensen, K., 123(15), 128
Jensson, O., 189(41), 191(41), 195
Jérôme, H., 78(28), 81(28), 97, 139(26), 141(26), 146
Jeune, M., 122, 131
Job, J. -C., 178(72), 180(68), 197, 198
Johnson, B. E., 139(41), 141, 148
Joly, C., 78(44, 48), 82(44), 83 (48), 98
Jonasson, J., 187(20), 193
Jones, H. W., 103(102), 135, 153 (19), 164
Jones, K. W., 152(24), 165
Jongbloet, P. H., 34, 36(48), 39, 66
Joss, E., 56(113), 71
Juberg, R. C., 78(22), 81, 96
Judge, C. G., 43, 66

K

Kadotani, T., 119(85), 124(85), 133
Kahlstrom, E. J., 186(27), 194
Kahn, J., 205(5), 207
Kaijser, K., 143, 146, 161(37), 166
Kajii, T., 36(53), 39(53), 53, 66, 67
Kaličanin, P., 78(33), 82(33), 97, 109(79), 133

Kaminetsky, H. A., 157(31), 165
Käosaar, M., 182(92), 185(92), 186 (92), 199
Kaplan, A. R., 108(49), 131
Kaplan, M., 185(51), 196
Kaplan, S., 47(112), 55(112), 71
Kardjiev, L., 202(1), 205(1), 207
Kasahara, S., 139(34), 140(34), 141(33, 34), 147, 203, 209
Kaselaid, V. L., 159(43), 166
Kask, V. A., 161(56), 167
Kattine, A. A., 183(55), 196
Kaufman, B. N., 153(19), 164
Kavanagh, T. M., 81(30), 97
Kawarazaki, T., 53(47), 66
Keay, A. J., 31(5), 32
Kelch, R. P., 140(33), 143, 146
Kemp, N. H., 157(39), 166
Kempen, C. van, 34, 36(48), 39, 66, 91, 96
Kennedy, J. L. Jr., 203(8), 207
Kerckvoorde, J. van, 183(26), 194
Khuri, N., 73(4), 75(4), 75
Kiffer, B., 54(80), 69
Kim, H. J., 186(45), 195
Kinch, R. A. H., 31(10), 32
Kistenmacher, M. L., 74, 76, 78 (24), 82, 96
Kitzmiller, N., 49(20), 64
Kivalo, E., 121(3), 127
Kletzky, O., 103(48), 130
Klinge, T., 205(28), 209
Klinger, H. P., 41(68), 68, 78(2), 81(2), 91(2), 95, 161(37), 166, 182(53), 196
Klose, J., 55(119), 72
Kluyskens, F., 122(51), 131
Knight, L. A., 181, 196
Koch, G., 107(96), 134
Koch, R., 92(41), 98
Koivisto, M., 143, 146, 180(14), 193
Komlos, L., 103(48), 130
Konigsberg, U. R., 103(83), 106 (83), 133
Koons, A., 86(10), 95
Koulischer, L., 51, 55, 57(83), 66, 69
Kouseff, B., 186(45), 195
Kramer, S. L., 182(2), 192
Krmpotić, E., 142, 147
Kučerová, M., 140(25), 144, 146
Kühner, U., 62, 66
Kunze, D., 123(80), 133
Kunze, J., 120, 125, 131
Kunze-Mühl, E., 101(30), 103(30, 38), 104(30), 108(30), 119(30), 129, 130
Künzer, W., 51(7), 63, 111(110), 115(110), 135, 141(46), 148
Kushnick, T., 58(100), 70
Kuznetsova, L. I., 161(42), 166

L

Labro, J-B., 154(6), 163
Laća, Ž., 109(79), 133
Lachance, R., 203(12), 208
Ladda, R., 73, 75, 76, 86(25), 96, 201, 205, 208
Lafourcade, J., 44(59), 46(60), 47 (58), 49(59), 50(56, 60), 51(57), 67, 78(28), 81(28), 97, 111(63), 114(63), 131, 132, 139(26), 141 (26), 146, 161(47), 166, 172(71), 183(71), 198, 203(17), 208
Lahav, M., 29(4), 32
Lamit, J., 122(56), 131
Lamy, M., 34(33), 36(33), 46(33), 49(33), 65, 101(44, 101), 103 (101), 105(44, 101), 110(46), 111(46), 114(46), 121(45), 130, 135, 190(39), 195
Langmaid, H., 184(93), 199
LaPolla, J., 50(65), 68
Larget-Piet, L., 172(71), 183(71), 198
Larson, S. L., 29, 32
Latta, E., 187, 196

Lauras, B., 73(5), 74(5), 76
Laurén, K., 187(20), 193
Laurence, B. M., 181(54), 196
Laurent, C., 46, 47(52), 53, 66, 73(11), 74(11), 76, 85(26), 93, 96, 108, 122(56), 131, 152, 155, 165, 184(21), 185, 193, 196
Lausecker, C., 78(15), 83(15), 96
Lautmann, F., 203, 208
Lavagna, J., 107(4), 127
Law, E. M., 81(30), 97, 111(60, 61), 114, 116, 131
Lazjuk, G. I., 120, 132
Leao, J. C., 36(53, 54), 39, 55(96), 67, 70, 157, 165
Lebas, E., 84(12), 95
Leclerc, R., 121(34), 129
Legros, J., 44, 57, 67
Lehre, R. Jr., 78(27), 83(27), 97
Lehrke, R., 78(27), 83, 97
Leisti, J., 103(62), 109(62), 121(3), 127, 132
Leisti, S., 103(62), 109(62), 132
Lejeune, J., 44, 46, 47(58), 49, 49(27), 50, 51, 65, 67, 78(28, 44, 48), 81, 82(44), 83(48), 86 (45), 97, 98, 106, 111(63), 114, 116, 121, 131, 132, 139, 141, 146, 161(47), 166, 170(67), 172 (71), 173(70), 178(72), 180(68), 183(69, 71), 184(21, 70), 185, 187(20), 188(67), 193, 196, 197, 198, 202(13), 203, 203(13), 208
Lele, K. P., 85(29), 91, 97, 170(52), 190, 196
Le Loch, J., 78(44, 48), 82(44), 83 (48), 98
Léopold, Ph., 36(45), 39(45), 66
Lessell, S., 150, 153, 166
Lesser, R. L., 86(20), 93(20), 96
Levenson, J. E., 103(67), 108, 132
Lévêque, B., 121(45), 130
Levy-Silagy, J., 54(98), 70
Lewis, F. J. W., 103(28), 105, 129
Lichtenberger, M., 46, 55, 67
Lin, C. C., 185, 199
Lindenbaum, R. H., 36(62), 42, 67, 143, 146, 211(3), 212
Lindsten, J., 117(31), 129, 156, 161, 166, 181(33), 182, 187 (20), 193, 196
Lion, D. T., 47(107), 55(107), 71
Littlefield, J., 73(10), 75(10), 76, 86(25), 96, 201(16), 205(16), 208
Littlefield, J. W., 115(29), 123 (29), 129
Littlefield, L. G., 83(18), 96
London, D. R., 157, 166
Longin, B., 155(33), 165
Look, R. A., 61(101), 70
Lorber, J., 142, 145
Lord, P. M., 181, 196
Lother, K., 54, 68
Lotker, M., 103(48), 130
Louchet, E., 56(106), 71
Lozzio, C. B., 183, 196
Lubs, H. A., 30, 31, 32
Lucas, M., 46(15), 61(15), 64, 190(8), 192
Luchsinger, R., 56(113), 71
Luciani, J. M., 161, 166
Luers, Th., 82(6), 95, 156(7), 163
Luft, R., 161(37), 166
Lurie, I. W., 93(38), 97, 120, 132, 161(42), 166
Luriye, I. V., 92, 97
Luzzatti, L., 181(48), 196
Lynch, H. T., 108(49), 131
Lyngbye, T., 75(6), 76
Lyons, R. B., 91, 98

M

Macaulay, M. E., 36(78), 39(78), 69

McCathie, M., 110(27), 118(27), 129
McCracken, J. S., 52, 67
Machin, G. A., 29, 32
MacIntyre, M. N., 49(27), 50, 65, 68
Maclaverty, B., 143(31), 147
Maclean, N., 153(13), 157(13), 164
McDermott, A., 107, 132, 189(41), 191(41), 195
McDonald, L. T., 77(56), 80(56), 99, 119(107), 121(107), 135
MacDonald, P. A. C., 111(21), 115 (21), 128
McFarland, B. S., 46(22), 53(22), 64
McGavin, D. D. M., 47(64), 55, 68
MacGillivray, M. H., 161(10), 163
McGilvray, E., 41(68), 68, 81(2), 91(2), 95
McNamera, B. G. P., 182(37), 195
McNamara, D. G., 155(46), 159(46), 166
McRae, K. N., 103(103), 106(103), 135
Maganias, N. H., 204, 208
Magenis, R. E., 140(28), 143, 147
Mahoney, M. J., 139(17), 144(17), 146
Maillard, E., 123(23), 128, 161(15), 164, 181(19), 193
Malfatti, C., 174(15), 193
Mallet, R., 202(13), 203(13), 208
Malpuech, G., 103(70), 108, 109, 132
Mangold, H., 211(5), 212
Mann, J., 62, 68
Mann, J. D., 105(26), 115(105), 129, 135, 187(95), 200
Mannini, A., 174(78), 198
Mantle, D. J., 153(13), 157(13), 164
Marchal, G., 107(36), 130
Marie, J., 121(45), 130
Marimuthu, K. M., 73(10), 75(10), 76, 201(16), 205(16), 208
Maroteaux, P., 91(17), 96
Marshall, R., 107(36), 111(21), 115(21), 128
Martin, B., 180(63), 197
Martin, J., 108(104), 135
Masters, P. L., 182(60), 197
Masterson, J. G., 81, 97, 111(60, 61), 114, 116, 131
Matsui, I., 205(21), 208
Mattei, J. F., 159(26), 162(27), 165
Mattei, M. -G., 162(27), 165
Maurus, R., 57(83), 69
Mauuary, G., 109(35), 129
Mavalwala, J., 92(41), 98
Medioli Cavara, F., 180(33), 194
Meer, J. -J., 86(45), 98
Méhes, K., 174(90), 175(9), 190 (9), 192
Melin, K., 161, 166
Mella, G., 139(2), 141(2), 145
Melnyk, J., 77(56), 80(56), 99, 86(58), 92(58), 99, 119(107), 121(107), 135
Melville, M., 31(5), 32
Menon, R., 73(1), 75(1), 75
Mercer, R. D., 203, 208
Merrill, R. E., 46(22), 53(22), 55(103), 64, 71
Merritt, A. D., 122(84), 133
Metz, F., 179, 196
Michaels, D. L., 124, 132
Michard, J. P., 151(14), 152(14), 164
Michel, Cl. van, 44, 57, 67
Michel, M., 108(59), 122(56), 131
Michiels, J., 74, 76
Mickell, J., 187(88), 199
Migeon, B. R., 103(73), 106, 133
Mikelsaar, A. -V. N., 91, 97, 110,

(Mikelsaar, A. -V. N.)
116, 122, 133, 142, 147, 159, 161, 166, 182, 182(92), 185(92), 186(92), 190, 196, 199
Mikelsaar, R. -V. A., 159(43), 161 (42, 56), 166, 167, 182(92), 185 (92), 186(92), 199
Mikkclsen, M., 34(21, 40), 38, 51, 59(86), 64, 66, 69, 81, 97, 103 (53), 107, 117(55), 131, 144, 147 156, 166, 174(47), 182, 185 (44), 186(47), 189(41), 191(41), 195, 196, 197, 205(28), 205, 208, 209

Miller, D., 78(51), 81(51), 98
Miller, D. A., 39(67), 40(114), 41 (68), 46(69), 53(69), 58(114), 63, 68, 71, 78(2), 81(2), 91(2), 95
Miller, O. J., 39, 40(114), 41, 44 (8), 46, 53, 58(8, 114), 59(8), 63, 68, 71, 78(2), 81(2), 91(2), 95, 124(13), 128
Miller, R. W., 86, 96
Milunsky, A., 54, 68
Ming, P. M. L., 178(98), 200
Mittwoch, U., 157(39), 166
Monteverde, R., 55(9), 63
Moore, C. M., 74, 76
Moore, M. K., 73(14), 74(14), 76, 111(77), 117, 123, 133
Morič-Petrovič, S., 78(33), 82, 97, 109, 133
Morineaud, J. -P., 44(61), 55, 67
Morishima, A., 156(7), 163
Mortensen, E., 182(59), 197
Mortezai, M., 51(26), 65
Morton, H. G., 137(2), 138
Moszer, M., 56(4), 63, 190(8), 192
Mouriquand, Cl., 36(45), 39(45), 66
Moyer, F. G., 42, 68
Mozziconacci, P., 106(64), 132
Mulcahy, M. T., 182, 197
Muller, J., 92(37), 97
Muller, H., 143(44), 148, 174(9), 175(9), 190(9), 192
Murken, J. -D., 123, 133
Murphy, J. W., 141(33), 147
Mürset, G., 175(42), 186(42, 81), 195, 198

N

Nakagome, Y., 205, 208
Nance, W. E., 46(22), 53(22), 64, 111(81), 115, 123(106), 133, 135
Naydionova, M. M., 159(43), 166
Negus, L. D., 34(116), 36(116), 41(116), 72
Neimann, N., 78(34), 80, 97, 107 (36), 130
Neu, R. L., 36(53, 54), 39(53, 54), 55(96), 67, 70, 124, 133, 204(2), 207
Neuhäuser, G., 47(74), 54, 57, 68, 157, 166, 204, 208
Neurath, P., 73(10), 75(10), 76, 201(16), 205(16), 208
Nevin, N. C., 143, 147
Niebuhr, E., 44(76), 60, 61, 62, 68, 77, 79(35), 81, 84, 86, 97, 117(55), 131
Niederer, B. S., 119(11), 124(11), 135
Nielsen, J., 31, 32, 123(15), 128
Nitowski, H. M., 103(83), 106, 133
Nivelon, A., 73(11), 74(11), 76, 122(56), 131
Noël, B., 47, 57, 69, 85(26), 93 (26), 96, 174(61), 197
Noël, G., 152(34), 165
Nolan, T. B., 55(96), 70
Nora, J. J., 155, 159, 166
Northcutt, R. C., 161(181), 164
Norwood, T. H., 178, 197

O

Obermann, U., 84(19), 96
Ockey, C. H., 36(78), 39, 69
O'Grady, R. B., 93, 97
Oikawa, K., 53(47), 66
Okken, A., 118(40), 130
Olive, D., 54, 57, 69, 157(49), 167
Ooghe, M. J., 73(5), 74(5), 76
Opitz, J. M., 92, 97
Orbeli, D. I., 92, 93, 97
O'Reilly, J. N., 205(5), 207
Orye, E., 86(40), 93, 98, 122(51), 131, 154(30), 165
Osborne, R. A., 62(104), 71
Øster, J., 75(6), 76
Ottosen, J., 77, 79(35), 84, 86, 97
Ovellette, E. M., 74(2), 75
Overton, K., 140(28), 143(28), 147
Owen, L., 180, 197

P

Paci, A., 51(28), 65
Page, D., 115(29), 123(29), 129
Palmer, C. G., 122, 133
Palo, J., 121(3), 127
Pan, S., 187(88), 199
Pant, S. S., 74(2), 75
Paolini, P., 106(64), 132, 161, 166
Parker, C. E., 92, 98
Passarge, E., 40(3), 41, 47(82), 54, 63, 69
Paterson, C. R., 110(27), 118(27), 129
Patriarca, P. L., 124, 128
Patton, R. G., 152(59), 167
Pearce, P., 190(65), 197
Pearson, G., 115(105), 135, 187 (95), 200
Peltier, J. M., 181(19), 193
Penneau, M., 172(71), 183(71), 198
Penrose, L. S., 85(29), 91(29), 97
Perciaccante, R., 124(13), 128
Pergament, E., 119(85), 124, 134, 206, 209
Perheentupa, J., 103(62), 109(62), 132
Perona, G. P., 174(78), 198
Peters, A., 78(34), 80(34), 97
Petersen, G. B., 144(21), 146, 187(20), 193
Petit, P., 55(50), 57, 66, 69, 122, 123(9), 127, 134
Pettid, F., 108(49), 131
Peyresblanques, J., 155, 167, 204(27), 209
Pezzani, C., 180(33), 194
Pfeiffer, R. A., 34(85), 36(85), 39, 52, 69, 83(53), 99, 106, 123(11), 127, 134, 186, 179 (56), 196, 197, 205, 207
Pfitzer, P., 103(88), 134
Philip, J., 59, 69
Philippe, N., 108(59), 131
Picciano, D. J., 83, 98, 125, 134, 143, 147
Pierson, M., 54(80), 69, 157, 167
Pietra, G. C., 119(85), 124(85), 134
Pinçon, J. A., 108(59), 131
Pitt, D. B., 43(46), 66, 190, 197
Plachot, M., 137(3), 138
Podugolnikova, O. A., 107, 127, 190, 197
Polani, P. E., 34, 44(87), 69, 152(24), 156(38), 165, 166
Polivková, Z., 140(25), 144, 146
Pomeroy, J., 74(15), 76
Poncelet, M., 123(9), 127
Poncelet, R., 122, 134
Porsch, R., 33(118, 120), 34(120), 36(118, 120), 38(118, 120), 72
Poty, J., 203(12), 208
Poujol, J., 107(4), 127
Poulsen, H., 205(20), 208
Powers, H. O., 204(2), 207

Prader, A., 175(42), 186(42), 195
Prats, J., 47(2, 93), 56(2), 63, 70
Praud, E., 78(44, 48), 82(44), 83 (48), 98
Predescu, V., 158, 167
Preto, G., 34(92), 39(92), 52(91), 70, 180(18), 193
Prieur, M., 86(45), 98, 178(72), 198
Pritchard, J. G., 158(20), 164
Prod'hom, A., 153(51), 167
Proesmans, W., 155(8), 163
Pruett, R. C., 86, 86(25), 96, 98
Puck, M., 159(16), 164
Pujatti, G., 170(77), 180(77), 198
Punnett, H. H., 50, 69, 74, 76, 78(24), 82, 96, 103(5), 111(24), 115(24), 127, 128
Putnam, T. I., 109, 128
Puyau, F. A., 123(106), 135

Q

Quack, B., 46(77), 47(77), 57(77), 69, 174(61), 197
Quersin, Cl., 51, 66
Quigley, M., 86(10), 95
Quinodoz, J. M., 153, 167

R

Race, R. R., 156(38), 166
Rafferty, J. H., 62, 68
Raoul, O., 202(13), 203(13), 208
Rashad, M. N., 81(30), 97
Raspiller, A., 157(11), 163
Ratcliffe, S., 31(5), 32
Ray, M., 31(3), 32, 103(103), 106 (103), 135
Raynaud, E. J., 103(71), 108(70), 109(71), 132
Razanamparany, M., 206(25), 209
Reboa, E., 205(11), 207
Ree, M. J., 161, 167
Reese, A. B., 86, 99
Reichelt, W., 54, 69
Reinwein, H., 33(118, 120), 34 (120), 36(118, 120), 38(118, 120), 51(7), 52, 55(119), 63, 70, 72, 106, 111(110), 115 (110), 134, 135, 141(46), 148, 173(43), 178(4), 182(96), 183 (43), 192, 195, 200, 205(29), 209
Reisman, L. E., 78(7), 95, 139, 140(34), 141, 147, 153(29), 165, 203, 209
Reiss, J. A., 55(96), 70
Reny, A., 154(57), 157(11, 12), 163, 164, 167
Réthoré, M. -O., 51(57), 67, 73 (5), 74(5), 76, 78(28, 44, 48), 81(28), 82, 83(48), 86(45), 97, 98, 106(64), 111(63), 114(63), 116(65), 116, 121(66), 132, 134, 139(26), 141(26), 146, 161(47), 166, 170, 172, 173, 178, 180, 183, 184, 185(51), 188, 196, 197, 198, 203(17), 208
Rey, J., 142(11), 145
Revazov, A. A., 181, 198
Ribas-Mundo, M., 47(2, 93), 56 (2), 63, 70
Ricci, N., 39, 52, 70
Richard, J., 57(83), 69
Richards, B. W., 124, 134
Richardson, F., 73(4), 75(4), 75
Richter, C., 56(99), 70
Ricks, Ph. Jr., 157(31), 165
Ridler, M. A. C., 50(5), 63
Rieu, D., 142(11), 145
Rimoin, D. L., 140, 143, 144(43), 148
Ritter, H., 106(90), 134
Robert, J. M., 46(52), 47(52), 53,

(Robert, J. M.)
66, 184(21), 193
Robertson, J., 110(27), 118(27), 129, 137(2), 138
Robinow, M., 190(86), 199
Robinson, A., 117(97), 134, 159 (16), 164
Robinson, J. C., 139(2), 141(2), 145
Roca, M., 47(2, 93), 56(2), 63, 70
Roe, A. M., 46(15), 61(15), 64
Rohde, R. A., 52, 70, 182, 198
Rohmer, A., 54(98), 70, 144(39), 147
Roidot, M., 73(8), 74(8), 76
Romano, P. E., 93(36), 97
Rommel, J., 74(15), 76
Rorke, L. B., 119(111), 124(111), 135
Rosi, G., 180(29), 194
Rossier, A., 178(72), 198
Rothstein, T. B., 93(36), 97
Rott, H. D., 107(96), 110(37), 117 (37), 134, 170(76), 182, 183(69), 184(84), 185, 197, 198, 199
Röttinger, E., 141(5), 145
Rottino, A., 124(13), 128
Roubin, M., 184(93), 199
Roux, C., 62(109), 71
Rowe, P., 107(69), 132
Rowley, J. D., 206, 209
Royer, J., 152(34), 165
Royer, P., 110(46), 111(46), 114(46), 130
Ruch, J. -V., 54(98), 70, 174(6), 192, 204(3, 26), 207, 209
Ruddle, F. H., 30, 31, 32
Rudiger, R. A., 41(81), 69
Rumpler, Y., 54(98), 70, 174(6), 192, 203(3), 204, 206, 207, 209
Rundall, T. S., 78(51), 81(51), 98
Rundle, A. T., 124(91), 134
Rutten, F. J., 185(46), 195, 211,
(Rutten, F. J.)
212
Ruvelcaba, R. H. A., 103, 107, 134
Ryan, R. J., 157(31), 165

S

Sachsse, W., 47(95), 58, 70
Sakaguchi, S., 181(48), 196
Sakai, K., 141(12), 145
Salamanca, F., 83, 98
Saldaña-Garcia, P., 187(20), 193
Salmon, C., 34(33), 36(33), 46(33), 49(33), 65, 101(44, 101), 103 (101), 105(44, 101), 106(42), 110(46), 111(46), 114(46), 130, 135
Salmon, Ch., 91(17), 96, 121(45), 130, 185(51), 196
Salmon, D., 91(17), 96
Saltiel, H., 151(14), 152(14), 164
Salzer, G., 123(80), 133
Samuelson, G., 161, 166
Sanchez, O., 185, 200
Sandberg, A. A., 161(10), 163
Sander, C., 117(50), 131
Sander, L. Z., 117(50), 131
Sanger, R., 156(38), 166
Sanroman, C., 116, 134
Santesson, B., 170(7), 178(7), 192
Saraux, H., 36(24), 42(24), 46(24), 47(24), 64, 78(44, 48), 79(11), 82(44), 83, 85(11), 86(45), 88 (11), 95, 98, 110(32), 120, 129, 134
Sarto, G. E., 155, 162, 167
Sartori, A., 170(77), 180, 198
Sartori, E., 174(78), 198
Sato, H., 119(85), 123(85), 134
Sauvage, P., 144(39), 147
Savary, J. B., 60(18), 64
Savilahti, E., 103(62), 109(62), 132

Say, B., 142, 147, 155, 167
Scaife, N. S., 142, 147
Schaak, J. -C., 157(49), 167
Schachenmann, G., 174, 198
Schaefer, P., 56(4), 63
Schaller, A., 103(38), 130
Schamberger, U., 184(84), 199
Scheres, J. M. J. C., 185(46), 195, 211(4), 212
Schindeler, J. D., 143(9), 145
Schindler, A. M., 111(109), 114 (109), 135
Schinzel, A., 118, 134, 179, 186, 191, 198
Schlegel, R. J., 55, 70, 157(35), 165
Schleiermacher, E., 56(99), 70
Schmickel, R. D., 34(102), 38(102), 70, 140(23), 143(23), 146
Schmid, E., 141(5), 145
Schmid, W., 47(12), 56, 60, 64, 70, 118(95), 134, 174(78, 82), 179, 186(81), 191(80), 198, 199
Schmidt, E., 47(95), 58(95), 70
Schmidt, G., 205(28), 209
Schneegans, E., 54, 70
Schoeller, L., 115(12), 123(12), 127
Schoultz, B., 121(3), 127
Schroeder, H. -J., 56, 70
Schröeder, J., 143(24), 146
Schroeder, T. M., 56(99), 70, 141 (20), 146
Schröder, J., 180(14), 193
Schröter, R., 33(120), 34(120), 36 (120), 38(120), 72
Schubart, G., 115(12), 123(12), 127
Schuler, D., 73(7), 74(7), 76
Schulz, J., 142, 147
Schwanitz, G., 107, 110(37), 117 (37), 130, 134, 170(76), 182 (75), 184, 185(76), 198, 199, 211, 212
Schwartz, J. J., 58, 70
Schwartz, R. S., 115(29), 123(29), 129
Schwenk, J., 56(99), 70
Sebahoun, A., 159(26), 165
Sebaoun, M., 137(3), 138
Sedano, H. O., 61, 70
Sedláčková, E., 36(108), 40(108), 71
Sellyei, M., 73(7), 74(7), 76
Semmelmeyer, U., 211(5), 212
Senders, V., 187(88), 199
Senzer, N., 162, 167
Sergovich, F., 31, 32
Seringe, P., 44(59), 49(59), 67
Seringe, Ph., 178(72), 198
Serment, H., 159(26), 165
Serville, F., 204, 209
Shaw, M. W., 170, 179, 199
Shearin, D. B., 51, 65
Shepard, T. H., 156(21), 164
Shibata, K., 144, 147
Sidbury, J. B. Jr., 34(102), 38, 70
Siebers, J. -W., 182(96), 200
Siegel, A. E., 115(105), 135, 187 (95), 200
Silber, D. L., 55, 71
Silvey, G. L., 211(1), 212
Simon, H. A., 52, 69, 123(11), 127
Sinclair-Smith, B. C., 123(106), 135
Sindhvananda, N., 103(83), 106(83), 133
Singer, H., 47(74), 57(74), 68, 142, 147
Singer, J. D., 78(13), 80(13), 95
Singh, D. N., 62, 71
Sinha, A. K., 155(46), 159(46), 166
Sitska, M. E., 122(76), 133, 182 (58), 196
Skakkebaek, N. E., 182(59), 197

Slungaard, R., 92(37), 97
Smith, D. W., 41(34), 65
Smith, L. B., 47(82), 54(82), 69
Smith, M. E., 77(5), 78(5), 79(5), 83(5), 95
Smout, M. S., 31(10), 32
Solitare, G. B., 44(105), 56, 71
Solmon, C., 46(60), 50(60), 67
Sommer, A., 140(14), 143, 146
Soukup, S. W., 36(29), 43, 47(82), 54(82), 65, 69, 174(31), 190, 194, 199
Soulayrol, R., 56(106), 71
Soulie, P., 190(39), 195
Sparkes, R. S., 78(49), 80(49), 98, 103(67), 108(67), 132, 140(4), 143(44), 145, 148, 187(13), 193, 206(6), 207
Sperling, K., 82(6), 95
Stahl, A., 56, 71, 121(6), 127, 158(1), 159(26), 161(40), 163, 165, 166
Stalder, G. R., 105(10), 127, 142 (47), 148, 174(9), 175(9), 186, 190(9), 192, 199
Stallard, H. B., 85(29), 97
Stanchi, F., 174(15), 193
Stanescu, B., 74(12), 76
Staples, W. I., 50(65), 68
Stark, G. D., 182(35), 194
Steele, M. W., 44(8), 47(107), 55 (107), 58(8), 59(8), 63, 71, 187, 199
Stephan, E., 120(57), 125(57), 131
Sternberg, W. H., 78(55), 81, 99
Stevenson, A. C., 178(16), 193
Stewart, A., 124(91), 134
Stewart, J. M., 117, 124(72), 132, 134
Stimson, C. W., 156(5), 163
Stoll, C., 144, 147
Storm, D. F., 103(18), 120(18), 125(18), 128
Struck, H., 186(64), 197
Šubrt, I., 36(108), 40, 71, 109, 134, 187, 199
Sujansky, E., 186(45), 195
Summitt, R. L., 105, 135, 144(43), 148
Suomalainen, E., 121(3), 127
Surana, R. B., 180, 199
Sutherland, G. R., 43(46), 66, 182 (35), 190(65), 194, 197
Suzuki, Y., 142(12), 145
Svenningsen, N., 139(16), 141(16), 146
Syme, J., 31(5), 32
Szepetowski, G., 81(3), 95

T

Taft, P. D., 174(91), 199
Taillemite, J. -L., 62, 71
Takagi, N., 161(10), 163
Talvik, T. A., 110, 116, 122(76), 133, 142, 148, 159(43), 161, 166, 167, 182(58), 182, 185, 186, 196, 199
Tamborino, G., 178(25), 194
Tanghe, W., 170(5), 178(5), 192
Taracha, B., 184(103), 200
Taviu, C., 158(50), 167
Taylor, A., 140(8), 142, 145
Taylor, A. I., 36, 41, 44(110), 71, 93, 98, 153(3), 163
Taysi, K., 103(100), 135
Tenconi, R., 170(77), 180(77), 183(3), 192, 198
Ten Kate, L. P., 118(40), 130, 179(34), 194
Teplitz, R. L., 78(51), 81, 98
Ter Haar, B. G. A., 185(46), 195, 211(4), 212
Terzakis, T. A., 47(107), 55(107), 71
Thelen, T., 78(27), 83(27), 97
Therkelsen, A. J., 205, 209
Theiler, K., 85(8), 88, 92, 95
Thieffry, S., 34(33), 36(33), 46 (33), 49(33), 65, 101(44), 101,

(Thieffry, S.)
103(101), 105(44), 130, 135, 139(26), 141(26), 146
Thiriet, M., 46(77), 47(77), 57(77), 69
Thomas, Ch., 154, 167
Thomas, G. H., 73(4), 74(13), 75(4), 75, 76
Thompson, H., 91, 98
Thorburn, M. J., 139(41), 141, 148
Thovichit, S., 111(24), 115(24), 128
Thuline, H. C., 103, 107, 134
Tiepolo, L., 117(31), 129, 174(78), 181(23), 193, 198
Tinaztepe, K., 103(100), 135
Titus, J. L., 29, 32
Tolksdorf, M., 81, 83, 99, 120(57), 125(57), 131
Toews, H. A., 103(102), 135
Tompkins, R., 52, 70
Toni, G., 178(25), 194
Torres, F. G., 155(46), 159(46), 166
Towner, J. W., 34(116), 36(116), 41 (116), 72, 85(59), 86(58, 59), 92 (58), 93(59), 99, 103(33), 129, 179(100), 186(27), 194, 200
Tremblay, M., 51(26), 65
Tridon, P., 157(12), 164
Tsuboi, T., 123(15), 128
Tunçbilek, E., 142(35), 147, 155 (54), 167
Turleau, C., 184, 199
Turner, H. H., 149, 167
Turner, J. H., 47(112), 55, 71
Turpin, R., 44(59), 46(60), 47(58), 49(59), 50(56, 60), 67, 130(26), 141(26), 146
Tusques, J., 202(1), 205(1), 207
Tüür, S., 182(92), 185(92), 186(92), 199
Tygstrup, I., 59(86), 69

U

Uchida, I. A., 77(56), 99, 103, 106, 119(107), 121(107), 135, 185, 199
Umansky, I., 78(13), 80(13), 95

V

Vaillaud, J. C., 108, 135
Vaharu, T., 152, 167
Valdes-Dupena, M., 103(5), 127
Valdmanis, A., 105(26), 115, 129, 135, 187, 200
Valentine, G. H., 31(10), 32
Vandanabeete, B., 86(40), 93(40), 98
Varela, M. A., 78(55), 81, 99
Vassela, F., 56, 71
Venecia, G. de, 92(37), 97
Venema, W. J., 78(22), 81(22), 96
Ventimiglia, B., 34(92), 39(92), 52(91), 70
Vernant, P., 190(39), 195
Verresen, H., 43(25), 65, 155(8), 163, 170(5), 178(5), 183(26), 192, 194
Veslot, J., 73(8), 74(8), 76
Vest, M., 142(47), 148
Vestermark, S., 144, 147
Vialette, J., 44(59), 49(59), 67, 116(65), 132
Vianello, M. G., 105(11), 207
Vieira, H., 49(17), 64
Vigneron, C., 109(35), 129
Vinograd, J., 156(5), 163
Vischer, D., 56, 70
Vogel, W., 182, 200, 205, 209
Voigt, G., 54(89), 69
Volpe, J., 139(2), 141(2), 145
Voorhess, M. L., 152(59), 157 (35), 165, 167

Vorsanova, S. G., 181(73), 198
Vougier, M., 158(1), 163

W

Wadia, R. P., 115(8), 127
Wahrman, J., 191, 200
Wald, S., 123, 135
Waldenmaier, C., 144(37), 147
Wallace, C., 205, 209
Walz, D. D., 60, 64
Walzer, S., 30, 31, 32, 178, 200
Wang, H. C., 77, 80, 99, 103(103), 106(103), 119, 121, 135
Warburg, M., 182, 200
Warburton, D., 39(67), 40, 41(68), 44(8), 46(69), 53(69), 58, 58(8), 59(8), 63, 71, 78(2), 81(2), 91 (2), 95
Warner, S., 78(13), 80(13), 95
Warren, R. J., 140, 143, 143(9), 144, 145, 148
Watanabe, G. -I., 142(12), 145
Watanabe, N., 124(82), 133
Waxman, S. H., 103(2), 127, 156 (21), 164
Weber, F. M., 143, 148
Weinberg, T., 108(83), 106(83), 133
Weiner, S., 190(65), 197
Weiskopf, B., 211(1), 212
Weiss, L., 111(24), 115(24), 128
Weleber, R. G., 140(28), 141(19), 142, 143(28), 146, 147, 148
Wertelecki, W., 111, 111(109), 114, 117, 135
Wesson, M. E., 155, 167
Wharton, B. A., 190(11), 192
Whyte, R., 110(27), 118(27), 129
Widmer, R., 186(87), 199
Wieczorek, V., 184(84), 199
Wiedemann, H. -R., 81(54), 83(53), 99
Wiener, S., 86, 99
Wiesmann, U., 56(113), 71
Wijffels, J. C. H. M., 51, 66
Wilcock, A. R., 36(115), 41, 72
Williams, C. E., 178(16), 193
Williams, J. D., 77(5), 78(5), 79(5), 83(5), 95
Williamson, C., 31(3), 32
Wilson, M. G., 34(116), 36(116), 41, 72, 78(9), 82, 85, 86, 86(59), 92, 93, 95, 99, 103 (33), 129, 179, 186(27), 194, 200
Winnacker, J. L., 204(18), 208
Wiscovitch, R. A., 62(104), 71
Wright, S. W., 78(49), 80(49), 86(10), 95, 98, 187(13), 193
Wolf, U., 33, 34, 36(118), 38, 51(7), 52, 55, 63, 70, 72, 106(90), 111(110), 115, 134, 135, 141, 148, 178(4), 192
Wolkowicz, M. W., 57(79), 69
Wurster, D., 74, 76

Y

Yamak, B., 142(35), 147
Yamamoto, M., 142(12), 145, 183(28), 194
Yanagisawa, S., 181, 200
Yanoff, M., 119, 124, 135
Yarema, W., 190(86), 199
Young, R. B., 180(22), 193
Yssing, M., 182(59), 197
Yunis, J., 103(39), 104(39), 107(39), 130, 185, 200
Yusuf, A. F. M., 158(20), 164

Z

Zackai, E. H., 77(5), 78(5), 79 (5), 83(5), 95

Zahnd, G., 153(51), 167
Zang, K. D., 47(74), 57(74), 68
Zaremba, J., 124(91), 134, 184, 200
Zdansky, R., 142, 148
Zdzienicka, E., 184(103), 200
Zellweger, H., 47(121), 55, 72
Zera, M., 205(11), 207
Zernahle, K., 54(89), 69
Zetterqvist, P., 117(31), 129, 182(53), 196
Zucker, E. I., 159(43), 166

SUBJECT INDEX

A

Abortions:
  recurrent and chromosomal translocation, 202
  spontaneous, 29
Albinism:
  in chromosomal translocation, 202
  in X chromosome abnormalities, 151
Amblyopia (see Refractive error)
Amniocentesis, 6
Anal atresia, in "cat-eye" syndrome, 174
Anaphase, 3-5
Aneuploidy, 17
Aneusomie de recombinaison, 23
Aniridia and Wilm's tumour syndrome, 73, 201
Anterior chamber cleavage syndrome, in chromosome 4 deletion, 36
Antimongolism, 139-140
Antimongoloid slant:
  in aneuploid translocations, 211
  in B group deficiencies, 34, 46
  in "cat-eye" syndrome, 174
  in C group duplications, 172
  in D group deficiencies, 78, 85, 88
  in E group deficiencies, 103
  in F group deficiencies, 137
(Antimongoloid slant)
  in G group deficiencies, 139, 140
Arhinencephaly:
  in D group deficiencies, 77
  in E group deficiencies, 101
  in trisomy 13, 104
Astigmatism (see Refractive errors)
Asymmetry (facial):
  in B group deficiencies, 34
  in C group deficiencies, 73
  in D group deficiencies, 85
Autoradiography, 10
  in diagnosis of B group deletions, 48

B

Banding, 7-13
  fluorescent, 10-12
  G-bands, 12
  Giemsa, 12-13
  Q-bands, 11-12
  R-bands, 12-13
Barr body, 25-26
Bivalent, 3, 5
Blepharocholasis, in G group deficiencies, 139-140
Brushfield spots:
  in B group deficiencies, 36, 47
  in G group deficiencies, 140

C

Cardiac abnormalities:
  in "cat-eye" syndrome, 174
  in D group deficiencies, 77, 85
  in E group deficiencies, 119
  in G group deficiencies, 140
  in Turner syndrome, 150
Cataract:
  in B group deficiencies, 33, 36, 46
  in "cat-eye" syndrome, 174
  in D group deficiencies, 79, 88
  in E group deficiencies, 119
  in G group deficiencies, 140
  in spontaneous abortions, 29
  in trisomy 13, 88
  in X chromosome abnormalities, 150
Cebocephaly, in E group deficiencies, 101, 103
Cell division, 1-5
  anaphase, 3-5
  interphase, 1, 3-5
  meiosis, 3-5
  metaphase, 3-5
  mitosis, 1, 3
  prophase, 3-5
  telophase, 3-5
Centric fusion, 23-24
Centromere, 1, 3, 5-6, 8
Chiasma, 5
Chicago classification, 6-7
Choroid (see Coloboma)
Chromatid, 1, 3, 6
Chromatin-negative, 150
Chromatin-positive, 150
Chromosomal abnormalities:
  incidence, 29-30
  in newborn infants, 30
  numerical anomalies, 17
  in perinatal deaths, 29
  in spontaneous abortions, 29
  structural changes, 17-24
Chromosome, 1-26
  acrocentric, 6-7
  A group, 7
    duplications of, 170
  banding, 7-9
  B group, 7
    deficiencies of, 33-62
    duplications of, 170-171
  centromere, 1, 2, 5-6, 8
  C group, 7
    deficiencies of, 73
    duplications of, 171-174
  Chicago classification, 6-7
  chromatid, 1, 3, 6
  classification and nomenclature, 6-9
  deletion (see Deletion)
  D group, 7
    deficiencies of, 77-94
    duplications of, 174
    rings, 77-79
  diploid number, 3
  duplications of, 176-177
  E group, 7
    deficiencies of, 101-126
    duplications of, 175
  examination of, 5-6
  F group, 7
    deficiencies of, 137
  G group, 7
    deficiencies of, 139-144
  haploid number, 3, 5
  identification, 10-16
  insertion (see Insertion)
  inversion (see Inversion)
  isochromosome, 18-20
  metacentric, 6-7
  mosaic, 17
  Paris classification, 7-9
  ring (see Ring chromosome)
  sex, 7, 24-26
  submetacentric, 6-7
  translocation (see Translocation)

(Chromosome)
X, 24-26
Y, 24
Coloboma (or iris, choroid, uveal tract):
in B group deficiencies, 33-37, 46, 48
in "cat-eye" syndrome, 174
in D group deficiencies, 78, 85
in E group deficiencies, 110, 119
in trisomy 13, 88
in trisomy 22, 140
Colour vision, in X chromosome abnormalities, 151
Corneal abnormalities:
in B group deficiencies, 36
in E group deficiencies, 103
in X chromosome abnormalities, 151
Cri-du-chat syndrome, 44-62
incidence, 44
ocular abnormalities, 45-48
phenotype, 44
Crossing-over, 3, 5
Cyclopia:
and abnormal karyotype, 103-104, 140
in spontaneous abortions, 29

D

Deletion, 17-19
Deoxyribonucleic acid, 1, 10, 46

E

Ears:
abnormally shaped:
in B group deficiencies, 44
in D group deficiencies, 77, 85, 86
in E group deficiencies, 101
(Ears)
low set:
in B group deficiencies, 34, 44
in B group duplications, 170
in E group deficiencies, 119
in G group deficiencies, 140
protruding, in C group duplications, 172
Enophthalmos:
in C group duplications, 172
in E group deficiencies, 110
Epicanthus:
in aneuploid translocations, 211
in B group deficiencies, 34, 44, 46-47
in C group deficiencies, 73
in D group deficiencies, 78, 85, 88
in E group deficiencies, 102, 110, 119
in E group duplications, 175
in F group deficiencies, 137
in G group deficiencies, 140
in X chromosome abnormalities, 150
Exopthalmos, in B group deficiencies, 36, 48

F

Facial asymmetry (see Asymmetry)
Fingers (see also Thumbs):
clinodactyly, in B group deficiencies, 34
malformed, in E group deficiencies, 101
tapering, in E group deficiencies, 110

G

Gametes, 3
G-bands, 12

Giemsa stain, 12-13
Glaucoma:
congenital, in chromosomal translocation, 202
in X chromosome abnormalities, 151

H

Hallermann-Streiff syndrome, 36
Hypertelorism:
in aneuploid translocations, 211
in B group deficiencies, 33-34, 44, 46-47
in "cat-eye" syndrome, 174
in C group deficiencies, 73
in C group duplications, 172
in D group deficiencies, 78, 85, 88
in E group deficiencies, 102, 110, 119
in X chromosome abnormalities, 150
Hypospadias:
in B group deficiencies, 34
in D group deficiencies, 85

I

Incidence:
of chromosome 4 deletions, 34
of chromosome 5 deletions, 44
of Turner syndrome, 149
Insertion, 17, 20-21
Interphase, 1, 3-5
Inversion, 17-19
paracentric, 18-19, 189, 202
pericentric, 18-19, 189
Iris (see Coloboma)
Isochromosome, 18-20

K

Karyotype, 13-16
evolution of, 176
numerical changes in, 17
structural changes in, 17-24

L

Lens (see also Cataract):
opalescent, in chromosome 4 deletion, 46
subluxation of, in spontaneous abortions, 29
Lid lag, in chromosome 4 deletion, 34
Lyon hypothesis, 25-26

M

Macular abnormalities, cherry-red spot, in D group deficiencies, 78
Meiosis, 3-5
anaphase, 4-5
metaphase, 4-5
prophase, 3-5
telophase, 4-5
Mental retardation:
in B group deficiencies, 34, 44
in B group duplications, 170
in "cat-eye" syndrome, 174
in C group duplications, 172
in chromosomal translocations, 202
in D group deficiencies, 77, 85-86
in E group deficiencies, 101, 110, 119
in G group deficiencies, 140
Metaphase, 3-5

Microcephaly:
in aneuploid translocations, 211
in B group deficiencies, 34, 46
in B group duplications, 170
in C group deficiencies, 73
in C group duplications, 172
in D group deficiencies, 77, 86
in E group deficiencies, 110, 119
in F group deficiencies, 137
Micrognathia:
in B group deficiencies, 34, 44
in D group deficiencies, 77, 85
in G group deficiencies, 140
Microphthalmos:
in B group deficiencies, 36
in C group deficiencies, 73
in D group deficiencies, 78, 85
in E group deficiencies, 119
in F group deficiencies, 137
in spontaneous abortions, 29
in trisomy 13, 88
Mitosis, 1, 3
anaphase, 3
metaphase, 3
prophase, 3
telophase, 3
Mixoploidy, 17
Mongoloid slant:
in B group deficiencies, 47
in D group deficiencies, 78, 88
in E group deficiencies, 103
Monosomy, 17, 23
of G group chromosomes, 139-140
Mouth:
carp-shaped, in E group deficiency, 110
downward slanting, in C group duplication, 172
Myopia (see Refractive errors)

N

Neck:
short:
in B group duplications, 170
in D group deficiencies, 85
in E group deficiencies, 101
webbing, in Turner syndrome, 149
Nose:
broad nasal bridge, in G group deficiencies, 140
broad root, 34
flat, in E group deficiencies, 101
in B group duplications, 170
in C group duplications, 172
prominent nasal bridge, in D group deficiencies, 85
Nystagmus:
in B group deficiencies, 33, 36, 46
in "cat-eye" syndrome, 174
in D group deficiencies, 78, 88
in E group deficiencies, 103, 110, 119
in E group duplications, 175
in X chromosome abnormalities, 151

O

Optic atrophy:
in B group deficiencies, 46
in "cat-eye" syndrome, 174
in E group deficiencies, 110-110, 120
in X chromosome abnormalities, 151
Optic nerve, staphyloma, in spontaneous abortions, 29
Ovarian dysgenesis (see Turner syndrome)

P

Paris classification, 7-9
Persistent hyaloid artery, in D group duplications, 175
Persistent tunica vasculosa lentis:
in B group deficiencies, 36
in D group deficiencies, 79
in E group deficiencies, 119
Polyploidy, 17
Position effect, 201
Prophase, 3-5
Ptosis:
in B group deficiencies, 33-34, 48
in D group deficiencies, 78, 85
in E group deficiencies, 102
in G group deficiencies, 140
in X chromosome abnormalities, 150
Pupillary abnormalities:
eccentric pupil in X chromosome abnormalities, 151
ectopia pupillae in B group deficiencies, 33, 36
fixed pupil in B group deficiencies, 36

Q

Q-bands, 11-12

R

R-bands, 12-13
Refractive errors:
in aneuploid translocation, 211
in E group deficiencies, 103, 111
Retinal abnormalities (see also Retinoblastoma and Retinal dysplasia):
(Retinal abnormalities)
in B group deficiencies, 46
in E group deficiencies, 103
in X chromosome abnormalities, 151
Retinal dysplasia:
in "cat-eye" syndrome, 174
in E group deficiencies, 119
in spontaneous abortions, 29
in trisomy 13, 88
Retinoblastoma:
in D group deficiencies, 79, 85-88
with abnormal karyotype, 86
Ring chromosome, 19
Robertsonian translocation, 22-24

S

Scalp defects, in B group deficiencies, 34
Scleras, blue, in X chromosome abnormalities, 150
Segregation, 21-23
Sex chromatin, 24-25
Strabismus:
in aneuploid translocation, 211
in B group deficiencies, 34, 35, 44, 46
in "cat-eye" syndrome, 174
in D group deficiencies, 78, 85
in E group deficiencies, 102, 110, 119
in E group duplications, 175
in F group deficiencies, 137
in X chromosome abnormalities, 150
Syndrome:
aniridia and Wilm's tumour, 73, 201
anterior chamber cleavage, 36
"cat-eye", 174

(Syndrome)
cri-du-chat, 44-62
Hallermann-Streiff, 36
Turner, 25-26, 149-162
Wolf-Hirschhorn, 33-43

T

Telomere, 19
Telophase, 3-5
Thumbs, absent or hypoplastic in D group deficiencies, 77, 85-86
Toes:
malformed:
in B group deficiencies, 34
in D group deficiencies, 85
in E group deficiencies, 101
Translocation, 17, 20-24
aneuploid, 211
balanced, 201-202
centric fusion, 23-24
insertion, 20-21
parental, balanced, 47, 103, 111, 170, 177
reciprocal, 20-21
Robertsonian, 22-24, 176
shifts, 20
Trigonencephaly, in D group deficiencies, 77
Trisomy, 17-23
D group:
13 compared with partial monosomy 13, 88
(Trisomy)
13 with cyclopia, 103-104
E group, 18 type and countertype, 120
G group:
21 compared with partial monosomy 21, 139-140
22 compared with partial monosomy 22, 140
Turner syndrome, 25-26, 149-162
incidence, 149
ocular abnormalities, 150-151
phenotype, 149

W

Wolf-Hirschhorn syndrome, 33-43
incidence, 34
ocular abnormalities, 34-37
phenotype, 34

X

X chromosome abnormalities (see Turner syndrome)
X-linked disease, 24-25

Z

Zygote, 3

DATE DUE

DEC 1 1 1985			
NOV 2 8 1995			
GAYLORD			PRINTED IN U.S.A.